John Eyton Bickersteth Mayor

**Modicus cibi medicus sibi or Nature her own Physician**

John Eyton Bickersteth Mayor

**Modicus cibi medicus sibi or Nature her own Physician**

ISBN/EAN: 9783337030698

Printed in Europe, USA, Canada, Australia, Japan

Cover: Foto ©berggeist007 / pixelio.de

More available books at **www.hansebooks.com**

# MEDICUS CIBI MEDICUS SIBI

OR

## *NATURE HER OWN PHYSICIAN*

BY

## JOHN E. B. MAYOR M.A.

**FELLOW OF ST. JOHN'S COLLEGE**
**AND PROFESSOR OF LATIN IN THE UNIVERSITY OF CAMBRIDGE**

Cambridge
MACMILLAN & CO.
1880

λέγω ὑμῖν μὴ μεριμνᾶτε τῇ ψυχῇ ὑμῶν, τί φάγητε ἢ τί πίητε

MATT. vi 25.

ὑπωπιάζω μου τὸ σῶμα καὶ δουλαγωγῶ

1 COR. ix 27.

ἔστιν δὲ πορισμὸς μέγας ἡ εὐσέβεια μετὰ αὐταρκείας

1 TIM. vi 6.

quod si quis vera vitam ratione gubernet,
divitiae grandes homini sunt vivere parce
aequo animo ; neque enim est umquam penuria parvi

LUCRETIUS V 1117—9.

But is there yet no other way, besides
These painful passages, how we may come
To death, and mix with our connatural dust ?
" There is" said Michael "if thou well observe
The rule of NOT TOO MUCH, by temperance taught
In what thou eat'st and drink'st, seeking from thence
Due nourishment, not gluttonous delight,
Till many years over thy head return :
So may'st thou live, till like ripe fruit thou drop
Into thy mother's lap, or be with ease
Gather'd, not harshly pluckt, for death mature"

MILTON, P. L. xi 527-537.

# Then shall they fast in those days

## ST. JOHN'S

*Second Sunday in Lent, 1880*

νήπιοι, οὐδὲ ἴσασιν, ὅσῳ πλέον ἥμισυ παντός,
οὐδ' ὅσον ἐν μαλάχῃ τε καὶ ἀσφοδέλῳ μέγ' ὄνειαρ

HESIOD *op. et d.* 40, 41.

οὐδ' ὑγιείης τῆς περὶ σῶμ' ἀμέλειαν ἔχειν χρή,
ἀλλὰ ποτοῦ τε μέτρον καὶ σίτου γυμνασίων τε
ποιεῖσθαι· μέτρον δὲ λέγω τόδ', ὃ μή σ' ἀνιήσει

*aur. carm.* 32-34.

κοινὴν δέ τινα συμβουλὴν ἄπασι τοῖς ταῦτα ἀναγνωσομένοις,
ἰδιώταις μὲν τῆς ἰατρικῆς, οὐκ ἀγυμνάστοις δὲ τὸν λογισμόν, ὑπο-
τίθεμαι τοιῶσδε· μή, καθάπερ οἱ πολλοὶ τῶν ἀνθρώπων ὡς ἄλογα
ζῷα διαιτῶνται, καὶ αὐτοὺς οὕτως ἔχειν, ἀλλὰ καὶ διὰ τῆς πείρας
κρίνειν, τίνα μὲν αὐτοὺς ἐδέσματά τε καὶ πόματα βλάπτει, τίνες δὲ
καὶ πόσαι κινήσεις   .     .     .     .     .
οὗτοι μὲν οὖν σπάνιοι καθ' ἑκάτερον τὸ γένος, οἵ τε μεγάλως
βλαπτόμενοι καὶ οἱ μηδὲν ἀδικούμενοι· τὸ δὲ μεταξὺ πᾶν ἐν τῷ
μᾶλλόν τε καὶ ἧττον εἰς τὸ πυλὺ τῶν ἀνθρώπων ἐκτέταται πλῆθος.
ὧν τοῖς πεπαιδευμένοις (οὐ γὰρ οἱ τυχόντες γε ταῦτα ἀναγνώσον-
ται) συμβουλεύω παραφυλάττειν, ὑπὸ τίνων ὠφελοῦνται καὶ βλάπ-
τονται· συμβήσεται γὰρ οὕτως αὐτοῖς εἰς ὀλίγα δεῖσθαι τῶν ἰατρῶν,
μέχρις ἂν ὑγιαίνωσιν

GALEN *de sanitate tuenda* vi 14 f. (vi 449, 450 Kühn).

ταχὺ γὰρ καταπίπτουσιν ἐπὶ τὸ δρᾶν τὸ μὴ ἐξὸν οἱ πάντα δρῶν-
τες ἃ ἐξόν.... εἰ γὰρ καὶ τὰ μάλιστα ἕνεκεν τῶν ἀνθρώπων ἐγένετο
τὰ πάντα, ἀλλ' οὐ πᾶσι χρῆσθαι καλόν, ἀλλ' οὐδὲ ἀεί

CLEM. AL. *paed.* II 1 § 14.

To the Members of the Cambridge University Branch of the Church of England Temperance Society.

IF I could have taken any pledge, beyond the baptismal vow, to bind my future life, I would have pledged myself never to publish a sermon. But when asked to print this Lent discourse for distribution among my hearers, I resolved not (as on former occasions) to let it appear *als Manuscript gedruckt*—with a tacit understanding that it was not to be placed in a public library, or published in any other way—but to give it to the world. From childhood I had known (having lived as a Vegetarian of the strictest observance one Lent when twelve or thirteen years of age) that men may enjoy perfect health on a diet far more spare than is common among Englishmen of any class. By degrees I had approached a natural diet in my own personal practice, *e.g.* tobacco I have never smoked or chewed; from the time I came to college I had meat but once a day as a rule; butter, tea, coffee, strong wines, have for many years been to me rare luxuries; cocoa and light wines have followed in their train, and in my rooms whole-meal bread, porridge and water, have been my staple fare. As a proselyte to the Vegetarian society I have added fruit to *farinacea* and so become more luxurious.

Yet from sheer ignorance and sloth I have conformed to culinary customs which in my heart I have condemned as wasteful and irrational. Thus I am to blame, as much as any one, for the popular superstition which ranks fellows of colleges with aldermen as lovers of 'good things.' A guest of mine contrasts the plain living of Roger Ascham's days with the 'bloated luxury' of our present Cambridge.[1]

When I was a child, few Englishmen of any class took meat (at least in any quantity) more than once a day. The three heavy dinners (called breakfast, lunch, dinner) which ruin the health of the wealthy now and maintain an army of quacks, as yet were not. I remember many years ago being alarmed by the ignorance (I think) of Lord Clarendon, who, as a member of the Public Schools Commission, asked every headmaster whether his boys had meat three times a day. Nowadays philanthropists, in institutions professing to help poor students, find it hard to make both ends meet, because forsooth brain work requires a 'generous' diet of meat and strong drink. So entirely forgotten are the laws of health taught and practised by saints and sages of all time.

Plutarch (VII sap. conv. 16 p. 160*b*, Holland's

translation, p. 341) observes: "our eating and drinking is not only the meanes of our life, but also the cause of our death: for thereupon a number of diseases take hold of our bodies, which . . . proceed, no lesse from fulnesse than emptinesse, and many times we have more adoe to concoct, consume, and dissipate our food, than we had to get and provide it. And much like as if the daughters of *Danaus* were in doubt what to do, and what life to lead, or how to be emploied, after they were delivered and freed once from their servile task imposed upon them, for to fille their tunne boared full of holes; even so doubt we (in case we were come to this passe, as to cease from stuffing and cramming this unsatiable flesh of ours, which will never say Ho, with all sorts of viands that land or sea may affoord) what we should do? and al because for want of experience and knowledge what things be good and honest, we love all our life time to seeke for to be provided of necessaries: and like as they who have beene slaves a long time, after they come once to be delivered from servitude, do of themselves and for themselves the very same services, which they were woont to performe for their masters, when they were bound; even so, the soule taketh now great paines and travel to feed the bodie, but if once she might be dispatched and discharged from this yoke of bondage, no sooner shall she finde herselfe free and at libertie, but she will nourish and regard her-

selfe, she will have an eie then to the knowledge of
the truth, and nothing shall plucke her away, or di-
vert and withdraw her from it."

Suppose that Cambridge, this May term, set itself
to teach plain living, by precept and example, to its
gay visitors : no longer *spending money for that which
is not bread, and labour for that which satisfieth not.*
Suppose that our missionaries, instead of 'committing
the great blunder of throwing too much drink and
too much meat in the face of the Indian nation,'[2]
taught our countrymen there that free living in a
tropical climate is death. Suppose the clergy and
schoolmasters, who go out from among us, had sat
(like George Herbert and Nicholas Ferrar) in the
school of Cornaro and Lessius;[3] what would hap-
pen ?

Listen to the physician who has done more than
any living man to make the laws of health popular,
whose voice is always raised on behalf of temperance,
chastity and mercy. Dr. Nichols tells us (Herald of
Health, Mar. 1880, report of his first lecture in Cambridge, 30 Jan.) : "There is no reason why students
should be compelled to eat a luxurious and unhealthy
diet of flesh, or even to pay for it.[4] Every head would
be clearer, the blood of every student would be purer,
if they would live on sixpence a day, and they would
study none the worse if they were obliged to earn it.
A university ought to be a seat of learning and not

of luxury, and there is room for great reforms not only at Cambridge and Oxford, but at Eton and Harrow. The Scottish students who carried their sacks of oatmeal to Edinburgh, Aberdeen or Glasgow, were not the worse students for their homely and frugal fare, and there is better brain food in a basin of oatmeal porridge, than can be found in sirloins of beef or legs of mutton. . . . If the young men in schools and universities would live temperately,—live like refined Athenians, if not like the more hardy Spartans,—we should not see so many bloated and gouty aristocrats, nor so many ruined constitutions, among those who are foolish enough to follow bad examples."

It would be well if our faculty of medicine, of which we are justly proud, and which is now on the point of being endowed at the expense of the colleges, following the example of Dr. Nichols and Dr. B. W. Richardson, would teach laymen the meaning of the golden rule μηδὲν ἄγαν in respect of meats and drinks. Surely they might make it impossible for any resident in Cambridge to have recourse to quacks. Meanwhile, such of us laymen as have made a rule of life for ourselves (and I for one, having never regulated my diet by professional 'order,' have not known illness for forty-five years) may encourage others also to follow Galen's advice,[5] *i.e.* to learn by personal experiment what diet suits their constitutions. Plutarch (de sanit. 24 p. 136*e*, Holland, p. 626) " *Tiberius*

*Cæsar* was wont to say: That a man being once above three-score years of age deserveth to be mocked and derided, if he put forth his hand unto the physician for to have his pulse felt. For mine own part, I take this speech of his to bee somewhat too proud and insolent; but me thinks this should be true: That every man ought to know the particularities and properties of his own pulse, for there bee many diversities and differences in each one of us: also that it behooveth no man to be ignorant in the severall complexion of his owne bodie, as well in heat as in drinesse: also to be skilfull what things be good for him, and what be hurtfull, when he useth them: for he that would learne these particularities of any other than of himselfe, or goeth to a physician to know of him, whether he be better in health in summer time than in winter; or whether hee stand better affected in taking 'dry things rather than moist; also whether naturally he have a strong pulse or a weake, a quicke or a slow; surely hath no sense or feeling of himselfe, but is as it were deafe and blinde, a stranger he is dwelling in a borrowed body, and none of his owne: for such points as those are good to be knowen and easie to be learned; for that we may make proofe thereof every hower, as having the body with us continually.

"Also meet it is, among meats and drinks, to know those rather which be good and holsome for the

stomack, than such as be pleasant for the tooth;[6] and
to have experience of that which doth the stomacke
good, more than of that which is offensive thereto; as
also of those things that do not trouble and hinder
concoction, than which content and tickle the taste.
For to demand of a physician, what is easie of digestion, and what not; what doth loose, and what
bindeth the belly; me thinks is no lesse shamefull
than to aske him, what is sweet, what bitter, what
soure, tart or austere. But now we shall have many
folke, that know well how to find fault with their
cooks and dressers of meat for seasoning their broths
or making sauce to their viands, being able to dis-
cerne which is sweeter than it ought to be; which is
over-tart or too much salted: and yet they themselves
are not able to say whether that which is put into the
bodie and united therewith be light or no; and
whether it be harmlesse, not offensive, or profitable.
Hereupon it is that their pottage misseth not often
the right seasoning; whereas contrariwise, for want of
well seasoning their owne selves, but daily faulting
therein, they make much worke for physicians: for
they esteeme not that pottage best which is the
sweetest, but they mingle therewith many sharpe
juices and soure herbs, to make it somewhat tart
withall; but contrariwise, they send into the bodie all
maner of sweet and pleasant things, even untill it cry,
Ho; partly being ignorant, and in part not calling to

minde and remembrance that nature adjoineth alwaies
unto things that be good and holsome, a pleasure not
mingled with displeasure and repentance. Moreover,
we are likewise to remember and beare in minde all
those things that be fit and agreeable to the bodie, or
contrariwise, in the changes of the seasons in the yere,
in the qualities and properties of the aire, and other
circumstances, to know how to accommodat and apply
our diet accordingly: for as touching all the offences
proceeding from nigardise, avarice and pinching, which
the common sort doe incurre about the painfull inning
and laborious bestowing or laying up of their corne
and fruits; who by their long watchings, by their
running and trudging to and fro, discover and bewray
what is within the bodie, rotten, faulty and ulcerous :
we are not to feare, that such accidents will befall to
learned persons or students, nor yet to states-men and
polititians, unto whom principally I have addressed
this discourse; but they ought to beware and eschue
another kinde of more eager covetousnesse and illi-
berall nigardise in matter of studie and literature,
forcing them to neglect and not regard their owne
poore bodies, which oftentimes being so travelled and
outwearied, that they can doe them no more service,
yet they spare them never the more, nor give them
leave to be refreshed and gather up their crummes
againe; but force that which is fraile and mortall to
labour a vie with the soule, which is immortall; that

(I say) which is earthly, to hold out with the spirit, that is heavenly. Well, the ox said unto the camell his fellow servant, who would not ease him a little of his burden: Thou wilt not helpe me now to beare somewhat of my charge; but shortly thou shalt carie all that I carie, and me besides: which fell out so indeed, when the ox died under his burden: semblably it hapneth to the soule, which will not allow the sillie bodie (wearied and tired) some little time of rest and repose: for soone after comes a fever, headach, dizzinesse of the braine, with a dizzinesse of the sight, which will compell her to lay aside all books, to abandon all good letters, disputations and studie; and in the end is driven to languish and lie sicke in bed together with it for company."

Thus we are called by high authority to be physicians to ourselves, and the vast experience accumulated by the Vegetarian society proves that it is feasible and easy, in every rank of society, even in this luxurious age, to live the life, as far as diet goes, of Socrates, or Curius, or St. Paul. Thousands have tried (to speak with a paper-hanger, Vegetarian Messenger, 1, 1851, suppl. p. 18) 'the new system of living without doctors or doctors' bills, and without butchers or butchers' bills.' It is self-discipline to which we are invited, and plainly we are at liberty to be a law unto ourselves.

The case is different when we are urged to uphold

penal laws, affecting the health or happiness of other living creatures, of woman, or child, or 'the dumb animals,' whose impotence under torture is eloquent in the ears of mercy.

Dr. Andrew Clark, addressing the students of the London Hospital in October 1876 (T. L. Nichols, Herald of Health, November 1876, p. 132) protested against the law which regulates vivisection : he trusts *that every member of this great profession, and every thoughtful man beyond its pale, will make this cause his own, and will offer to threatenings of fresh legislation such a united, earnest and implacable opposition that the statute-book of England shall never again be sullied by penal enactments against the just liberties of men. The highest heritage of humanity is in our keeping. All the past and all the future conspire to make us loyal to the sacred charge, and at whatsoever cost of whatsoever kind we must hand down the freedom of experimental inquiry unmortgaged to future generations.* The comments of Dr. Nichols (*l.c.*) will prove that the medical profession is not unanimous in this view of 'the just liberties of men.'

Some years ago I met at Basel an enthusiastic young German physiologist. He complained that vivisection of the human subject was as yet forbidden, but looked forward to a millennium of science, when these shackles would be removed. He had probably never read Celsus, or he would have known that there was once a golden age of free science, but Celsus hugged his chains.

Vivisection of the human subject, which is now only a 'frommer Wunsch,' was possible to the ancients : (Celsus I praef. p. 4 l. 35 Daremberg) the dietetics taught *necessarium . . . esse incidere corpora mortuorum, eorumque viscera atque intestina scrutari ; longeque optime fecisse Herophilum et Erasistratum, qui nocentes homines, a regibus ex carcere acceptos, vivos inciderint, considerarintque etiamnum spiritu remanente, ea quae natura ante clausisset.* Some condemned the practice as cruel (p. 7 l. 15) *neque esse crudele, sicut plerique proponunt, hominum nocentium, et horum quoque paucorum, suppliciis remedia populis innocentibus saeculorum omnium quaeri.* The empirics regarded such torture as misleading no less than cruel (p. 7 l. 14) *atque ea quidem, de quibus est dictum, supervacua esse tantummodo ; id vero, quod restat, etiam crudele : vivorum hominum alvum atque praecordia incidi et salutis humanae praesidem artem non solum pestem alicui, sed hanc etiam atrocissimam, inferre ; cum praesertim ex iis, quae tanta violentia quaerantur, alia non possint omnino cognosci, alia possint etiam sine scelere ; (l. 34) ita mortui demum praecordia et viscus omne in conspectum latrocinantis medici dari necesse est tale, quale mortui sit, non quale vivi fuit. si quid tamen sit, quod adhuc spirante homine conspectui subiciatur, id saepe casum afferre curantibus. interdum enim gladiatorem in harena vel militem in acie vel viatorem a latronibus exceptum sic vulnerari, ut eius interior aliqua pars aperiatur, et in alio alia: ita sedem positum ordinem figuram similiaque alia cognoscere prudentem medicum, non caedem sed sanitatem molientem; idque per misericordiam discere, quod alii dira crudelitate cognorint. ob haec ne mortuorum quidem lacerationem necessariam esse, quae, etsi non crudelis, tamen foeda sit: cum aliter pleraque in mortuis se habeant, quantum vero in vivis cognosci potest, ipsa curatio ostendat.* Celsus himself (p. 12 l. 35) endorsed this censure: *incidere autem vivorum corpora et crudele et supervacuum est: mortuorum, discentibus necessarium: nam positum et ordinem nosse debent; quae cadavera melius, quam vivus et vulneratus homo, repraesentant. sed et cetera, quae modo in vivis cognosci possunt, in ipsis curationibus vulneratorum paulo tardius, sed aliquanto mitius, usus ipse monstrabit.*

In the early days of the Royal Society, while bear-baiting and bull-baiting were still in fashion, all Cambridge was of one mind with Dr. A. Clark. Barrow exclaims ('oratio ad academicos in comitiis' in his opuscula 128–9 or his works, Camb. ed. IX 46): *quin et oculos auriculis succenturiatis ac duci rationi comitem adiungitis experientiam. quando enim, obsecro, a condita academia in tot canum piscium volucrumque neces ac lanienas sanguinolenta curiositas saeviit, quo vobis partium constitutio et usus in animalibus innotesceret?* o innocentissimam crudelitatem et feritatem facile excusandam! So a Cambridge scholar, 15 Sept. 1648 (Sir T. Browne's works, 1836, I 360): *I have now by the frequency of living and dead dissections of dogs run through the whole body of anatomy.* Of Matt. Robinson, elected fellow of St. John's 3 Apr. 1650, we are told (Life, Camb. 1856, pp. 31–2) *in anatomy he was the most exquisite inquirist of his time, ... insomuch that he was invited by some learned persons in other colleges many years his senior to shew them vivisections of dogs and suchlike creatures in their chambers, to whom he shewed the whole history of the circulation, the* venae lacteae, *the cutting of the recurrent veins in the neck, with many experiments then novel, to great satisfaction, and no augur ever was more familiar with bowels than he: every week having some singularity or other of this nature to search in. Insomuch that one morning having been busy in his chamber with anatomising a dog, and coming to dinner into the college hall, a dog there smelling the steams of his murdered companion upon his clothes, accosted him with such an unusual bawling in the hall that all the boys fell a laughing, perceiving what he had been a doing, which put him to the blush.*

In the Menagiana, Amst. 1713, II pp. LII–LIII is an amusing squib on the 'old philosophy' (in a 'requête à nosseigneurs de Mont Parnasse'): *Que le sang ne circulera plus, et que le coeur ne lui ouvrira plus la porte pour entrer au poulmon. Que le foye sera réintegré dans son premier office de faire le sang, sans que le coeur lui ose plus disputer ledit office, et que le chile l'ira trouver tout droit par la veine porte sans s'amuser à aller monter vers les jugulaires, nonobstant aussi les oppositions experimentales de M. Pecquet,* auquel il sera nouvellement fait inhibitions et défenses de plus à l'avenir faire ouverture des chiens vivans pour prouver le contraire.

Has anything occurred since the seventeenth century to moderate raptures like Barrow's? Or is every 'thinker' bound to echo the war-cry of Dr. A. Clark?

Such books as Mr. E. B. Nicholson's 'Rights of an animal' (Lond. 1879)—a recent lecture on those rights by Prof. Chandler at Oxford—these and other symptoms prove that in "the hell of animals" conscience begins to own a duty to the lower creation. Assuredly rank, even the highest, will not long screen the heroes of *battues* and the like cruelties from prosecution.

Few of us perhaps regret that our law prohibits entertainments such as those at which Matthew Robinson played the augur. With Juvenal[7] we see only degradation to woman in dallying with torture. But we may well doubt our competence to form an opinion on Dr. A. Clark's invitation. I have therefore sought professional advice. I will call the writer X, because I am forbidden to make the name public. It is not safe, in this nineteenth century, for physicians to proclaim opinions counter to the fashion of the hour.

"It is certainly possible," writes X, "as my own experience shews, to pass all the . . . examinations required . . . without even once witnessing a vivisection; but it is impossible to escape studying these cruel experiments as recounted in the various books one has to 'get up' for the examinations. I have several times been asked the method, results and inductions of vivisectional experiments, and have of course been compelled to reply. This fact however does not hinder me from asserting that whatever knowledge may have been attained by such means, could have been otherwise obtained in nine cases out of ten, and I do not consider that the tenth exceptional case compensates fairly from a scientific point of view for the mass of error and false induction to which the practice of vivisection has undoubtedly given rise. The obscurity surrounding the study of the localisation of the various motor centres of the brain is a good example of misleading tendency of vivisectional experiments. [Then follows a full explanation of this point.] It is clear to the student of nervous disease, that such experiments cannot have any real value, and that slow as may be the progress of knowledge acquired by clinical observation, it is far better to wait for the development of such observations than to rush incontinently and impatiently to false and obscure conclusions obtained by such experiments as Ferrier's. . . . .

"Of course, from a moral point of view, which is the only real standard of vision for a civilised person, vivisection is absolutely barbarous and abominable, no matter what may or could be expected from it. For my own part, I prefer to take my stand on the moral ground entirely, for if once one admits the *principle* involved in vivisection as a legitimate one, I do not see where one is to draw the line. If experiments on animals can be admitted as allowable, because certain physiologists conceive that they may serve great ends in science, why not admit experimentation upon infants, lunatics, paupers and various

other comparatively low grades of the human race? Remember
that 'science' would necessarily profit far more by such experiments than by those conducted on mere animals, and remember
also, that the men who devise and perform vivisections recognise
no difference whatever between humanity and the brute creation.
For them, 'as one dieth, so dieth the other;' every thing, in
their creed, is soulless, irresponsible, ephemeral, and the mere
outward aspect of form is all that divides the man from the dog.
I fail then to perceive why they should respect the one more
than the other.

"For myself I hold a belief, I should say a 'knowledge' of a
far different kind, for in my creed all life is eternal, progressive,
responsible, and the 'incorruptible Image of God' is in all
creatures, 'shining more and more unto the perfect day' as its
expression becomes more and more perfectly human and Christ-
like."

In a tract (entitled 'vivisection') issued by the
international association for the total suppression of
vivisection (25 Cockspur Street, Charing Cross), we
read :

"Man is man, in our view, chiefly because he can discern
good from evil, not because he is a cleverer kind of monkey than
other monkeys, or because he can recollect more facts and put
them to better practical use than creatures in a lower stage of
development. Humanity is, therefore, a word of which we fully
accept the popular definition, and for us a man is human in proportion as he is humane. We do not admit a torturer to be a
man ; he is simply an individual of the genus Simia—an intelli-
gent individual if you like, but he has nothing human about
him. And when one of these animals says that 'cruelty is
necessary,' it sounds in our ears precisely as if he had said,
'robbery is necessary' or 'deceit is necessary,' or any other
habit of the lower grades which humanity has outgrown.

"We have just witnessed in Paris an unparalleled spectacle,

the incongruity of which would be ridiculous if it did not also
furnish melancholy evidence of the lack of understanding and
thought prevalent in a nation which claims to rank among the
most civilised in Europe. I refer to the part taken by M. Paul
Bert, the most notorious vivisector of the day, in the discussion
upon M. Ferry's Bill. What can be said of a state of manners
which permits such a man as M. Paul Bert to pose as a
moralist before the public—a man whose whole career has been
one long course of cruelties so varied and appalling that even
here, under the shadow of the Ecole de Médecine itself, they
have attracted special comment and associated the name of their
perpetrator with all the worst of the barbarities of a fallen
science? This Paul Bert, who appears now before Paris as the
champion of morals, is the same who, at the Exposition last
year, exhibited pictures of dogs undergoing the agonies of
tetanos induced by the administration of various poisons at his
hands,—pictures, the public display of which excited expressions
of censure and disgust in the columns of a well-known Parisian
journal. This is the same too, whose laboratory is the scene of
such awful horrors that persons living near the waste grounds
surrounding it have more than once complained to the authorities
of the shrieks and groans issuing from its walls, and even now,
while I write these lines, the Parisian law courts are occupied
with an action brought against this man by the proprietor of a
neighbouring hotel for loss of *clientèle* and other grievances,
caused by the continual howling and cries of the dogs 'used' in
his experiments.

"What better terms can be found to characterise the work of
Paul Bert's own life than the words he himself used in the
Chamber of Deputies: 'Such things as these, and such a method
of teaching as this, inspire indignation and disgust ; they are
like a bog in which one treads in mire !'

"Paul Bert is himself one of the most distinguished of Jesuits,
for he adopts in theory and carries into practice daily their
distinctive doctrine, 'The end justifies the means,' and, in common with all vivisectors, he argues that 'cruelty is necessary,'

that good may be obtained by evil, and that private and professional motives sanctify the perpetration of deeds which, if committed by the vulgar outside the profession, would be highly reprehensible, and punishable by law. In the view of these priests of materialism, public opinion has no right to set moral limits to the pursuit of material science; knowledge, no matter how attained, is the one positive and good thing, and morality, being a mere question of national habit, is entitled to secondary consideration only, if, indeed, to any consideration at all.

"By common consent, however, mankind, more truly inspired, recognises as its highest ideal of development One whose greatness was not owing to scholastic learning or to retentive memory, but to those very attributes which materialistic experts (I will not call them 'philosophers') regard as derogatory and unbecoming in an age of enlightenment; attributes such as mercy, gentleness, love, patience, sympathy with suffering and the like; in fact, to the identical qualities which they label in a bundle as 'sentiment,' and thrust aside with contempt.

"Are we to go back to our monkey ancestors then, and relinquish all the advantages we have gained, and for which we have toiled so hard and endured so much since the anthropolithic days of Haeckel? God forbid! The manhood in this English nation protests, and will not protest in vain, against the attempt which is now being made upon national morality by formulating into a legal principle the axiom that might is right. For man is man, not because he is a strong beast or a supremely sagacious beast, but because he has it in him to know and to love justice and to refrain from doing evil. And to such an one the plea that a method involving the torture of others is a right method because it has proved useful in the attainment of knowledge, carries no weight whatever. Is there any class of crime or any depth of baseness for which the same plea may not be urged? Does not falsehood sometimes appear useful to liars, and may not violence, fraud, theft, or even murder find apologists on the same grounds? True, the policy of the liar, thief, or coward generally fails in the long run, and so also does that of the professional torturer. It is no secret that the practice of vivisection has given rise among scientists to dissensions, difficulties, and errors which are incessantly accumulating, and which have sown the paths of physiology with a fruitful crop of false deductions and bewildering contradictions. And if among the millions upon millions of cruel experiments on living animals by means of which science has been well-nigh arrested, and true progress hindered so disastrously, some few have accidentally proved of service in the elucidation of a nascent discovery, no proof exists that such discovery would not have been vouchsafed by more legitimate means, nor do such isolated cases atone in the smallest degree for all the agony, heart-hardening, and degradation of manhood which they entailed on the miserable victims and their more miserable tormentors.

"Vivisection useful? Cowardice useful? Deliberate devilry useful? Sir, we who are men will not buy knowledge at the cost of our manhood, we will not sell for so pitiful a mess of pottage the divine birthright of humanity. As to our physical health that is not called in question, for no one who has been medically educated will seriously assert that the science of healing is in any way related or indebted to the practice of physiological torture.

"I have received my own medical education at the Faculté of Médecine in Paris. At the Ecole, Professors Béclard, Vulpian, and others vivisect almost daily. It is no exaggeration to say that the walls of that Inferno re-echo from morn to sun-set with shrieks and cries and moans, the supreme pathos of which no pen can render. When first I heard them, now long ago, I took them for the cries of children under operation, so terribly human were they in expression and appeal. And now, whenever I go there, knowing what they are, these cries strike and tear my heart and move me to a passion of indignation which is all the more terrible to endure because it is so impotent.

"I ask myself and you, sir, by what right do vivisectors thus outrage me and other men, and why are they permitted to make life intolerable to their superiors? It is not only a question of torturing horses and dogs and rabbits, it is a question of torturing men and women. I am tortured, and thousands of human beings are tortured with me every day by the knowledge that this infamous practice is being carried on in our midst with impunity. For my own part—and I know but too well that I express the feeling of a large number of my countrymen—it is literally true that the whole of my life is embittered by the existence of this awful wrong. Since I have known that vivisection is, and how it is practised, I have moved, and slept, and eaten, and studied, under the shadow of it, and its effluvium has poisoned for me the very air of heaven.

"I appeal in my own name and the names of all those men and women whom the vivisectors are torturing with me,—I appeal to the English Parliament for personal relief and for example to the world, and I most earnestly press upon the members of both Houses not to regard this question as one having a merely technical or limited interest. The day on which England finally sweeps this curse of torture from her schools and affirms the principle that civilised man may not seek advantage for himself by means of the agony and tears of any creature whom God has made dependent on him, will be a day of mightier import to the advance of civilisation than any which has dawned since she, first of all nations, spoke the word which made free men of slaves through every land in Christendom.

"There were vested interests then, there are vested interests now. But she made no sordid compromises then, she stooped to no half-measures. She faced the outcry of opposition fearlessly, and she led the world. But now the old spirit seems wanting, and the only legislation she has dared to make on this new question of Right or Wrong is at once untenable and impotent. Here is an evil so base and so hideous that it has excited a national agitation, and the law, in order to satisfy the conscience of the country, restricts the perpetration of the offence to certain licensees. Why not treat burglary, arson, fraud, &c., in a similar manner? Either the practice is right or it is wrong. If right, interference is worse than impertinent; if wrong, it is as wrong for A as it is for B, and to license and protect the crime in A while condemning and punishing it in B is an insult to common sense, and an outrage on the most elementary principles of morals, of law, and of civilisation."

"Behind the scenes" writes in the Parisian, 4 March 1880 :

"I have no small claim to be heard in the matter, seeing that its theory and its bearings have almost exclusively occupied me for the past ten years, that I am one of the earliest agitators against vivisection in England, and that I have been medically educated at the Ecole de Médecine in this city. I am therefore, by study and by profession, probably better qualified to gauge the value of scientific torture than your correspondent at Naples, whom I take to be, in the view of the Faculty, a layman. Let me handle his two statements in order.

"1.—If the society or societies to which he alludes have recently directed their energies chiefly against the practice of vivisection, it is because their members have perceived two important points which totally escape your correspondent's acumen; first, that vivisection is the *worst* kind of cruelty extant, inasmuch as it constitutes an organised system, and claims for itself a legitimate existence which no other form of cruelty has assumed; and secondly, that it is logically inconsistent and impossible to punish minor barbarities in the lower classes while granting impunity to the infinitely worse atrocities of the so-called higher grades of society. Moralists have nothing to do with the motive for crime, they have to deal with the crime itself. Which is worse, the suffering inflicted on a horse by cruelly lashing it under a heavy load, or the suffering inflicted on the same creature by flaying it alive, dissecting out its nerves, and torturing them with hot irons or with corrosive acids?... Why is the moralist to respect the excuse of the vivisector rather than that of the carter? With what face is he to say to the poor carter, 'I have nothing to do with your circumstances or with your motive, you are guilty of a barbarous and revolting action, and you must go to prison,' while in the same breath he tells the vivisector, ' I am sorry for the torment of your victim, but your excuse is sufficient justification, proceed with your work'? Would not this be a crying example of the principle—One law for the rich, another for the poor ? . . . .

"2.—This brings me to the second erroneous statement made by your correspondent. He says that vivisections have been useful to science. I answer boldly that they have more disastrously hindered science, and more completely degraded it than it is possible to conceive. Witness the innumerable contradictions, the wilderness of differing theories with which the literature of physiology and pathology teems, due to nothing more nor less than to the multiplication of vivisectional experiments! The same experiment never gives exactly the same results in two different hands, and as exact results are necessary to the establishment of a theory, every vivisector has a theory of his own, on which he bases interpretations of clinical facts often utterly at variance with the results of the observations of practising physicians. And no wonder, for vivisectional experiments are not a serious method of investigation. If we share with animals the brotherhood of suffering, we yet differ widely from them in the higher and more subtle phenomena involved in the direct and reflex action of the nervous system ; the more widely, in fact, in proportion to the grade of mental and physical development we have attained. And the nervous system dominates, pervades, and controls all other systems and tissues of the body in such a degree that no phenomena occurring within the region of the vascular, muscular, cellular, or even the osseous systems, can be explained without intimate reference to the brain and spinal centres. This being so, no man will persuade me while I retain my senses that because certain effects have been observed by him in the healthy body of a dog or rabbit subjected to torture, and perhaps even 'curarized,' he can by that light interpret effects observable in the diseased body of a man, the springs and habits of whose nervous life and functions are so intimately different from that of the beast. I say that such a pretence as this is not serious nor scientific, and that consequently it is for the purposes of serious science worse than worthless—it is misleading.

"I know of no subject now occupying human thought, and the history of our epoch, which goes so thoroughly to the very root of philosophy, and involves so deeply the issues of moral sentiment, as this of vivisection, and I venture to assert, from my own knowledge of the question and of its bearings, that it is precisely on the platform occupied by this question that the coming battle between Materialism and Philosophy will have to be fought out."

In like manner Dr. Hogan, a London physician, sometime assistant 'in the laboratory of one of the greatest of living experimental physiologists' writes (Dietetic Reformer 1875, pp. 190–1)) :

"In that laboratory we sacrificed daily from one to three dogs, besides rabbits and other animals, and after four months' experience I am of opinion that not one of those experiments was justified or necessary. The idea of the good of humanity was simply out of the question. During three campaigns I have witnessed many harsh sights, but I think the saddest sight I ever witnessed was when the dogs were brought up from the cellar to the laboratory for sacrifice. Instead of appearing pleased with the change from darkness to light, they seemed seized with horror as soon as they smelt the air of the place, divining apparently their approaching fate. They would make friendly advances to each of the three or four persons present, and as far as eyes, ears, and tail could make a mute appeal for mercy eloquent, they tried it—in vain. Even when roughly grasped and thrown on the torture-trough, a low complaining whine at such treatment would be all the protest made, and they would continue to lick the hand which bound them till their mouths were fixed in the gag, and they would only flap their tail in the trough as their last means of exciting compassion. Often when convulsed by the pain of their torture this would be renewed, and they would be soothed instantly on receiving a few gentle pats. It was all the aid or comfort I could give them, and I gave it often. They seemed to think it an earnest of fellow-feeling that would cause their torture to come to an end—an end only brought by death. Were the feelings of experimental physiologists not blunted, they could not long continue the practice.

The Herald of Health (1 May 1880, p. 352) gives from the Anti-Vivisectionist, the following letter from Dr. R. S. Butcher, of Dublin, University lecturer on Operative Surgery, ex-President of the Royal College of Surgery of Ireland, &c., &c. :

"Dublin, Feb. 18, 1880. Sir,—In answer to your letter, I beg to state that I firmly believe that no advantages to science can follow the cruel and demoralising practice of vivisection. Such an exhibition must be a disgrace to the cold, heartless, and would-be scientific professor; and most detrimental to the gentle and kindly feelings of the class of students that should be compelled to witness this sad cruelty. I hope you may bring opinion to bear so powerfully as to blot out this stain upon human nature." This was also the opinion of the late eminent surgeon, Mr. William Fergusson, as it is the opinion of many other men of the highest rank in medicine.

This evidence, which might easily be multiplied, may suffice to shew that on this question[8] 'doctors disagree' so widely, that the laity cannot follow the rule *unicuique credendum est in arte sua*. The history of medicine denies to the physician an infallibility which Protestant churches no longer claim. How can we deliver heretics to the secular arm for compulsory cure, when (as in the case of vaccination[9]) high authorities regard the cure as worse than the disease, syphilis,[10] *e.g.* worse than small-pox?

The great triumphs of medicine in this age, as in every age,—the 'miracles of healing,'—have been works of faith, hope and love; witness the humanity which inspires the medical classics; witness names like Amalie von Lasaulx and (after every abatement has been made) Sister Dora; witness legions of devoted women exorcising evil spirits and their works from army and navy, from the military hospital and the penitentiary; witness martyrs of science and of mercy like Charles Murchison, with his self-chosen epitaph POST MORTEM VITA. Verily they who sigh for prayerless wards are fallen on evil times. In these true healers Hygieia is incorporate as an angel of light—temperate, sober, chaste, merciful: all men know her voice and follow her bidding. But if she comes as stern Necessity, with iron scourge and torturing hour, with pains and penalties, inquisitors and spies, intoxicants and opiates, she forfeits quietness and confidence which are her strength. And if acts arming her with exceptional powers, *ne quid res publica detrimenti capiat*, are smuggled incognito through the legislature, men will spurn her aid: ἐχθρῶν ἄδωρα δῶρα κοὐκ ὀνήσιμα. Can we conceive Galen promoting compulsory Vaccination or the Contagious Diseases Act (Women)?

Professor Newman's argument against medical infallibility is cogent (Dietetic Reformer 1869 108-9) :

"It does not rest with Parliament to enact how a disease shall
be treated. If a bill were proposed to enforce that every one
who is seized with apoplexy should be bled, the Lancet would
probably be foremost in the outcry. I should expect it to pro-
pound that Parliament is no authority in medicine ; that to
protect us from dangerous treatment by ignorant pretenders,
Parliament enacts medical degrees as mere tests of knowledge,
but it must not dictate to those who have displayed their knowledge by gaining the degree. Nor is it to the purpose to say that
Parliament took advice of physicians before it legislated. Some
thirty or forty years ago, when homœopaths first disused bleeding
for apoplexy and fever, the disapproval of their conduct by the
orthodox medical Faculty was so universal and so vehement that
Parliament might easily have got medical warrant to enforce
bleeding. Nay, one hundred years ago physicians were zealous
for inoculation. My father was with difficulty saved from it by
the sturdy refusal of his mother, who said (as she told me) 'If
God send small-pox on my child, I must bear it ; but never will
I consent to give it to him on purpose : how could anyone know
what would come of it ?' At that time Parliament might have
been advised by eminent and learned physicians to make inoculation compulsory ; and I have no doubt those physicians spoke
as dogmatically to my grandmother in favour of it as any can
now speak of vaccination ; yet, by the advice of physicians,
inoculation is now made penal ! It is certainly possible that by
the advice of physicians vaccination also will hereafter be made
penal. Medicine is a changing, and (let us hope) progressive
Art; it has no pretension to be Science, or to have any fixed-
ness at all. The editor of the Lancet has probably read the
article in the Quarterly Review of April, 1869, entitled 'The
Aims of Modern Medicine.' It is a storehouse of detailed fact
for those who are too young to remember what it relates of unanimous medical error, pernicious on the hugest scale. Medi-

cine cannot improve unless the younger and fresher minds among physicians are left perfectly free to deviate from the routine of their elders. Nothing can justify Parliament in enacting a medical creed or enforcing any special medical procedures. But if physicians must have hands unfettered, have patients no right to choose their physician?—no right to repudiate treatment which they think quackery? We all ought to be re-vaccinated periodically, according to the Lancet. Does then Parliament dare to enact such a thing? It does not, else I might be taken by force and vaccinated to-morrow.........
One who carries disease with him is ostensibly dangerous. This—and this only—justifies legislation against him. But when a man or child is ostensibly healthy, no case is made out for legislation at all. To enact that a healthy person shall have a disease lest hereafter he get a worse disease, is a form of despotism hard to parallel ; and what is peculiarly disgraceful, it is directed against innocent infants alone, because they are helpless : it does not dare to attack us adults. This fact justly arouses parents to indignation. Let Parliament enact that every M.P. shall be at once vaccinated, and that it shall be done from arm to arm among them, every four or five years as the doctors may prefer, if they will enact such things concerning children. The law now says to a parent—'We are alarmed to see that your child has no disease. Cow-pox (for the public good) it must have, with the chance of other hideous diseases : submit, or else make yourself a criminal, have your hair cropped, and dress in prison garb.' Such legislation implies that Parliament is a Medical Pope, and would justify no end of monstrous violations of sacred personal right. The Lancet 'begs respectfully to tell me' that in the matter of 'vaccine lymph' 'the State (!) and private practitioners take great care.' Is this very comforting—very reassuring—to one who has read Ira Connell's frightful case? I have a paper before me—reprinted from the Lancet of Nov. 16 1861—which contains a detailed account of 46 children in Piedmont being infected with loathsome disease— soon fatal to some of them—from receiving the lymph (called

vaccine!) out of the arm of one child called (and supposed to be) healthy. As the surgeon cannot be omniscient, he cannot know the diseases hidden in a particular child; he is not to blame for not knowing; but this is precisely the reason why Parliament ought much rather to forbid than to enforce the vaccination of one child from another. It makes the enforcement so indefensible, that one is unwilling to affix the right epithet. But even if cows would kindly get cow-pox for our convenience, so that each child might have the disease direct from the cow, even so it would be blind tyranny for the law to say to a parent—'You shall not keep your child in perfect health; that is too dangerous a course.' When to this the parent replies by defiance of the law, and is treated as a criminal, the law-makers are (in my opinion) the real criminals before God and man. Parents who become martyrs by resisting the law, deserve a sympathy akin to those who are martyrs of religion."

---

Medicine, we have seen, speaks with two voices, the voice of invitation, *Venite;* and the voice of command, *Compelle eos intrare.* The first voice, modest and winning, makes us our own keepers : *Heal yourselves; shun the causes of disease; prove all things, hold fast that which is good; everywhere and in all things avoid excess; 'ne quid nimis.'*

The other voice, imperious and scornful, commanding what is not manifestly for our good—what has to many proved death, or ruin worse than death—dares not trust to persuasion or to the common laws; it claims special exemptions, and operates by fines and imprisonment. The more we obey the gentle monitress, finding her yoke easy and her service perfect freedom, the more deaf shall we be to the threats of the despot. The soberest men in England suffer nothing from Nature's laws, which they reverently observe; *per contra* they suffer most from man's artificial laws, which violate the :first laws of health. But they have their reward, for wherever they turn their eye they find evidence for that discipline by which they themselves keep a sound mind in a sound body. Primitive tribes, innocent of pathology, untainted by luxury, enjoy almost unbroken health. And where man begins, there he ends. *Numquam aliud natura, aliud sapientia dicit.* The ancient physicians, medical classics of all time, an increasing number of the leading physicians now, many 'laymen,' including the chief glories of our race, have taught that temperance—in a far stricter than the conventional acceptation—is the one source of health.

If we share this belief and wish to act upon it in life, we need not set any great machinery in motion, but by quiet exercise of personal liberty may, each in his sphere, help to set our neighbours free. If the prevalent excess brought happiness, or were the result of deliberate choice, one might despair of influencing others; but it is a burden, heavy to be borne, imposed by tradition on men's shoulders. I never met any one who thought our public dinners rational; if they were not sanctioned by tradition, no sober man, none, but a Nero or a Vitellius, would in cold blood

devise such prodigality of money and time and
health; none, but a misanthrope, fling such an apple
of Discord between class and class. How the public
health suffers may be read in the advertisements of
patent medicines, or in the number of druggists' shops.
Consider the very word 'pill.'

To pass from generals to our own academic com-
monwealth :—any student may satisfy himself that
money and time and vital force are squandered here
on eating and drinking. Witness every experienced
tutor, every medical adviser, every member of the
senate who finds college luxury a clog in the work
of his riper manhood.[11]

If any student has gone with me thus far, he (or
she, for the perils of excess touch both sexes) may
venture to apply a few simple rules. Assuredly it is a
great support (as we see in the history of the Church
and of religious orders) to attach oneself to a society
whose bond is temperance; but these rules may be
proved outside of any organisation. Members of all
branches of the Church will find what is here recommended included in the elements of their Christian
profession.

I PERSONAL EXPERIMENT. If fashion, medical
and lay, has for many years prescribed excess in
meats and drinks,[12] it is possible that you are guilty of
it. If you are dyspeptic, Nature is calling you to re-
pent. Examine your habits by the standard laid

down, *e.g.* by Dr. Nichols ('The diet cure' and 'How to live on sixpence a day'), or by prison dietaries. You will probably find that you eat and drink too often and too much, and that of too stimulating a quality. Diminish the amount and simplify the kind, until you can work as well after meals as before.

You must expect opposition. Society, politely blind to excess, is eagle-eyed to detect the heresy of abstinence. *A man's foes are they of his own household.* But if you are resolute and rational, criticism will only open the door for missionary work. Impatience in the devotees of tradition is a confession of weakness.

II NOT TOO MUCH. Having discovered what is enough to keep mind and body in full play, stick to your measure, as near as may be, always. We flout lord mayors and aldermen for their feasts; but we, the guests, are in truth the main sinners. The receiver is worse than the thief. The demand creates the supply. We, the public, impose this shameful tax on our rich men. Thirty men, abstemious as Pythagoras or Beda or Keshub Chunder Sen, scattered among the Mansion House company, would after a few trials, lighten the groaning board. Dishes untouched would soon vanish. I blush to think how late I learnt this simple lesson. It is no part of a guest's duty to eat or drink more than he wants; nor does he want more when in company than when alone. The son of Sirach says well (ecclus. xxxi 12, cf. the whole ch.): *If thou sit at a bountiful table, be not greedy upon it, and say not, There is much meat on it.* Hear too that perfect gentleman, George Herbert (who wrote, remember, when guests had much less freedom than now) :

Drink not the third glass,—which thou canst not tame,
When once it is within thee, but before
Mayst rule it as thou list,—and pour the shame,
Which it would pour on thee, upon the floor.
  It is most just to throw that on the ground,
  Which would throw me there if I keep the round.

   ·    ·    ·    ·    ·    ·

Shall I, to please another's wine-sprung mind,
Lose all mine own? God hath given me a measure
Short of his can and body; must I find
A pain in that wherein he finds a pleasure?
  Stay at the third glass; if thou loose thy hold,
  Then thou art modest, and the wine grows bold.

If reason move not gallants, quit the room—
All in a shipwreck shift their several way;
Let not a common ruin thee intomb:
Be not a beast in courtesy, but stay,—
  Stay at the third cup, or forego the place:
  Wine above all things doth God's stamp deface.
                   (*Church Porch* v vii viii).

Temperance is the true 'Moly that Hermes once to wise Ulysses gave,' sovereign charm against swinish excess even in the bowers of Circe.

III EXAMPLE. Temperance societies know that the abstinent alone can reclaim drunkards. So, be sure, none but the frugal can avert from our towns cycles of waste and want. Luxurious folk mutter that the 'working classes' are prejudiced against cheap and wholesome food. Yet a large proportion of the members of the Vegetarian society are 'working' men and women. Let our clergy learn *e.g.* what are wholesome weeds, and how to cook them; let them ask their people to a dinner of whole-meal bread and cooked weeds, where all share alike, and thousands of tons of valuable food may be saved, while the people's blood is purified. Leprosy, we have learnt, was banished from England with the advent of garden herbs. Let us once more take a lesson from France.

IV GIFTS. Never give, least of all to the young, what may breed wasteful and unwholesome habits. Tobacco and beer to boys, cast-off finery to girls, have times without number been the first stage to ruin.

A friend, now holding a high position in the university, has put in writing an anecdote which he told me many years ago. The pauper spirit would soon die out, if we all dealt as wisely with such illegal claims.

"The man, a journeyman carpenter, was sent by his master to do some repairs in my house, and I happened to come near the place where he was working. He at once stopped and requested me to give him a little beer. I said, 'Doesn't your master pay you enough wages to get beer for yourself when you need it?' He couldn't complain in the matter of wages, he got what other people got. 'But then you seem to want something more?' Well, all workmen asked for beer, he was only doing like the rest. 'Then you all need an advance in wages? Suppose we arrange it in this way. I don't like to give you beer, but you shall charge, in addition to your wages, what a glass of beer will be worth, or two glasses if you claim them in the day, and you shall tell your master to add that sum on my bill, and perhaps that will move him to give you what you want. Clearly he should pay you and not I; for suppose this were an empty house, there would be nobody here whom you could ask and so you would lose your beer, and if it is right you should have it, that would be unfair.'

"He shook his head, and said he thought it would be better not to trouble the master. Then said I, 'You have no real claim for beer, it seems? Now I wish English workmen would give up putting themselves in a position in which they are lowered, for it can never be satisfactory for you or anybody else to find himself in the position of a claimant for what is not his.' We parted very good friends, and afterwards the man, who became himself a master workman, stopped me one day, a Sunday morning when on his way to church, to thank me for our talk about beer-begging."

V RECIPROCITY IN ENTERTAINMENTS. Horace, the intimate of Maecenas and Augustus, knew well (c. III 29, Conington) :

"In change e'en luxury finds a zest :
The poor man's supper, neat, but spare,
With no gay couch to seat the guest,
Has smooth'd the rugged brow of care."

Thomas Walker[13] (The Original, no. 23, Oct. 21, 1835 II 279–282, ed. B. Jerrold, Lond. 1874) has admirable remarks on 'equality of style :'

"No one thinks of requiring an equality of sense, or wit, or learning, and why should the rule be different with respect to dishes or wines, except from the vulgar-minded feeling that money is more estimable than those qualities?. . . The sensible mode of proceeding is for all to keep regularly to that style which best suits their means, and then intercourse will find its true level. If the man of luxurious style seeks the society of his neighbour of simple style, it is because he finds some equivalent, and it is a loss to both that pride should bar their intercourse."

Robert Wilson Evans (The Bishopric of souls 1877, pp. 192–3) :

"The modest old parsonage is everywhere fast disappearing, and in its place rises a mansion which may enable its owner to entertain the highest company in the neighbourhood in the same way that he is entertained by them. . . . But formerly his old, inconvenient and straightened mansion was at once a check, and tied him down in some degree to its characteristic simplicity. His great friends, and such he had, enjoyed the change from their own luxurious style of living ; and they enjoyed it the more, because it was in harmony with the self-denying profession of a man of refinement and education."

If we are bound to return hospitality in kind and degree, 'save me from my friends' must be our motto. One spendthrift may devastate a whole society, to the profit of cooks and wine-merchants, and to the loss of house and home.

———————

Sallust remarks:[14] *Sovereignty is easily maintained by the qualities which won it at the first.* The Christian church is learning the truth of this maxim. A return to the primitive traditions, profounder study of the Bible, the revised translation (a work from which even R. W. Evans foreboded disunion among Protestants) have healed the breach between many leading Anglicans and the learned Nonconformists. And will not primitive life and conversation, the martyr-spirit going forth to grapple with vice in high and low, be attractive as of yore to more generous natures beyond the pale of any church? Still to this day, *Semen est sanguis Christianorum.* Unconscious Christians (Matt. xxv 34–40) were never more numerous than now, as Richard Rothe, a man of prophetic insight, proclaims with earnest and glad conviction. Professor Newman will teach us how the church, without compromise, rather by forswearing compromise, may win the world.

"When Woman is duly honoured and homes are purified,
And Fiery Drink is withheld from the weak in mind,
And the traffickers in Sin are pursued as Felons,
And Truth is open-mouthed, and Thought is Free;
God shall soon bless the land with blessings undreamed of.
Labour shall be honoured and enmity of classes cease,
Poverty shall be light-hearted, Pauperism shall wane,
Beggary and Roguery shall be trades extinct,
The jails and the houses of the Insane shall be idle,
Health shall be robuster and Orphanhood rare,
Orphans shall meet new love in families,
Youth shall be reared to pure thought, pure fancy,
High hope, high desire and tender piety.
Religion shall grow wise and Knowledge religious,
Atheism shall waste away, and Selfishness learn to blush,
And God shall be our God and we will be his people."

"But first of all must it be made a precept of Religion
And a precept of Politics, to root up *the causes* of Evil;
It must become a Creed, that debasement is unnatural,
Is therefore unnecessary, and is surely preventible;
That it is our duty to prevent, and will be our blessing;
That those who promote the body's welfare, aid the mind;
And that the Moral must precede the Spiritual, in national
      growth,
Though a few, out of immorality, be rescued into Spiritualism."

"Let us *in truth* heal ourselves, rising in the strength of God,
That strength which already abounds in the hearts of England;
Let the good join with *better* or *worse* to extirpate avowed evil,
And five years shall now do more than ever before did fifty,
And perhaps ere long men will doubt of Sin's vitality."

"Cleanliness and Health are conditions of general Virtue,
Conditions of Contentment, removing misery from Poverty.

Cleanliness and Health are the birthright of every savage :
Surely that 'civilisation' is barbarous which steals them from
    the poor.
Why should not Religion, now equally as of old,
Lift up her voice for every right of man,
And enforce duty on individuals, whether for body or mind ?
Man's conscience responds to every such faithful utterance,
Nor would the ministers of religion long protest in vain."

(Theism doctrinal and practical, 1858, pp. 2, 160, 162, 178.)

J. E. B. M.

*St. John's,* 6 *May* 1880.

-------

## L'ENVOI.

G O now with some daring drug
   Bait the disease, and while they tug,
Thou, to maintain their cruel strife,
Spend the dear treasure of thy life :
Go, take physic, doat upon
Some big-named composition,
The oraculous doctor's mystic bills,
Certain hard words made into pills ;
And what at length shall get by these?
Only a costlier disease.
Go, poor man, think what shall be

Remedy against thy remedy.
That which makes us have no need
Of physic, that's physic indeed.

Hark hither, reader, wouldst thou see
Nature her own physician be?
Wouldst see a man all his own wealth,
His own music, his own health?
A man, whose sober soul can tell
How to wear her garments well;
Her garments that upon her sit
(As garments should do) close and fit:
A well-clothed soul, that's not opprest
Nor chokt with what she should be drest?
Whose soul's sheathed in a crystal shrine,
Through which all her bright features shine,
As when a piece of wanton lawn,
A thin aerial vail is drawn,
O'er Beauty's face; seeming to hide,
More sweetly shews the blushing bride?
A soul, whose intellectual beams
No mists do mask, no lazy steams?
A happy soul, that all the way,
To heaven rides in a summer's day?
Wouldst see a man whose well-warmed blood
Bathes him in a genuine flood:
A man, whose tuned humours be
A set of rarest harmony?
Wouldst see blithe looks, fresh cheeks beguile
Age? wouldst see December smile?
Wouldst see a nest of roses grow
In a bed of reverend snow?
Warm thoughts, free spirits, flattering
Winter's self into a spring?

In sum, wouldst see a man that can
Live to be old, and still a man;
Whose latest and most leaden hours
Fall with soft wings, stuck with soft flowers:
And when life's sweet fable ends,
His soul and body part like friends:
No quarrels, murmurs, no delay;
A kiss, a sigh, and so away?
This rare one, reader, wouldst thou see?
Hark hither, and thyself be he.

RICHARD CRASHAW, *In praise of Lessius.*

I HATE when vice can bolt her arguments,
And virtue has no tongue to check her pride.
Impostor, do not charge most innocent Nature,
As if she would her children should be riotous
With her abundance: she, good cateress,
Means her provision only to the good,
That live according to her sober laws,
And holy dictate of spare temperance:
If every just man, that now pines with want,
Had but a moderate and beseeming share
Of that which lewdly-pampered luxury
Now heaps upon some few with vast excess,
Nature's full blessings would be well dispensed
In unsuperfluous even proportion,
And she no wit incumbered with her store;
And then the giver would be better thanked,
His praise due paid; for swinish gluttony
Ne'er looks to heaven amidst his gorgeous feast,
But with besotted base ingratitude
Crams, and blasphemes his feeder.

MILTON, *Comus* 760-779.

*" Then shall they fast in those days."*

ONE hundred and sixty-nine years ago our tutors sent to their pupils two discourses on the use, measure and manner of Christian fasting by Edward Brome, one of the fellows.[1]  One of these discourses is on our text as given by St. Mark.  They are dedicated to the master, Dr. Gower, then near his end,[2] as to 'a constant observer of church fasts.'  A saintly student, Ambrose Bonwicke,[3] whose tombstone lies mouldering in the neighbouring churchyard, notes that he had most of the book that was useful, in short, in Mr. Nelson, *i.e.* in the *Festivals and Fasts*, a book of great vogue.

Brome's preface complains of the general neglect of the holy church fasts, 'whereby too much occasion is given to our adversaries of the Roman communion . . . to publish it in Gath that in England there is no fasting.'

Fifty years earlier our text served as the theme of Dr. Gunning's discourse on the Lent fast.[4]  The author commended abstinence not only by precept, but by his looks, 'the most graceful and venerable,' says our historian,[5] 'I ever saw.' Gunning's monument in Ely cathedral fully bears out Baker's words. Two of his successors in the mastership have left characters of him. Humphrey Gower tells us:[6] 'Plenty of all things flowed round about him, but for the use of others rather than himself. His study and his business was his meat and drink, for of any other he had as little regard and made as little use, as it was well possible to flesh and blood. He that had writ so irrefragably for the fasts of the church, kept them as rigidly himself. But that suffic'd him not: He obliged himself to so many others, that they who knew not what excellent use his mind made of those hours of abstinence might suspect that so much severity to his body, inclin'd somewhat towards a fault.' 'He seem'd nothing more than a provident and faithful steward for the publick and the poor.' So Robert Jenkin, in his Latin character of Gunning,[7] testifies that he kept his gate and heart ever open to distress, stinting himself alone, lavish to others : *Afflictis semper et ianua et cor patuit, soli sibi parcus, sumptuosus aliis.*

Gunning's treatise was a reply to the presbyterian demand, 'that nothing should be in the

Liturgy, which so much as seems to countenance the observation of Lent as a religious fast.'[8] Not that Puritans condemned fasting as such; fasts appointed by themselves, *providential* fasts, they kept with great rigour, for example on the 25th of December.[9] But their hatred of Rome made them denounce all traditional fasts, as they did feasts (except the Sunday) and that most significant sign of the cross. When the Prayer-book was under ban, Gunning had bravely upheld the church's order: Cromwell's soldiers on Christmas day 1657 made prisoners of his congregation.[10] Hence when set free, he was the natural champion of the calendar.

If his arguments sometimes appeal only to one school in the church, as where he would impose the Lent fast as an apostolical tradition, in other points all Christians may well listen to this great master of the art.

Thus (249=495): 'The Christian law of liberty, which is not less binding because such, is principally a law of gratitude.' 'Fasting is the mother of health and a good habit of body; if thou believest not my word (it is Chrysostom[11] that speaks), ask the physicians about it and they will tell thee these things more clearly.' To this cause no doubt we must ascribe the fasts of

Augustus and Vespasian ;[12] they were kept by
medical advice. Another use of fasting[13] recognised by our catechism in the exposition
of the seventh commandment, is justly en-
forced again and again. Thus Leo the Great[14]
teaches (166=224) : 'Fasting hath ever been
the diet of virtue ; from abstinence do proceed chaste thoughts, reasonable wills, salutary
counsels.' Again, the connexion of fasting and
almsgiving is set forth by the same Leo (149
=200) : 'Then doth the medicine of fasting
work to the curing of the soul, when the
abstinence of him that fasts, refreshes the
indigence of him that hungers.' So the pro-
phet (Is. lviii 6-7) : 'Is not this the fast that
I have chosen,—to deal thy bread to the
hungry, and that thou bring the poor that
are cast out to thine house ? when thou seest
the naked that thou cover him ?' The highest
argument for Christian abstinence is stated in
Basil's words[15] : 'Our Lord having by fasting
fortified the flesh which He took for our sakes,
so received the assaults of the devil in it, in-
structing us by fastings to anoint and exercise
ourselves unto the combats of temptations.'
In brief, Gunning[16] teaches first what fasting is
not : viz. a gift to God for satisfaction to His

justice in lieu of eternal punishment; and secondly what it is: it is a judging of ourselves, that we may not be judged; it is a part of contrition and confession; it is for our future emendation and securing us from return to the same sin again, which hath caused us so to smart.

Time would fail to tell of our Puritan master, bp. Pilkington, part author of the homily on gluttony and drunkenness; of Dr. Wm. Lambe, a fellow of the college in the last century, from whom Abernethy learnt the medical efficacy of abstinence.[17] Passing beyond our walls, but still keeping within the university, hear bp. Grindal's words to Sir Wm. Cecil[18]: 'My opinion hath been long that in no one thing the adversary hath more advantage against us than in the matter of fast which we utterly neglect.' Or read Jeremy Taylor's house of feasting, apples of Sodom, Life of Christ[19]; read Isaac Barrow[20] on those words, 'Provide things honest in the sight of all men,' where (living under the second Charles) he paints with unrivalled wealth of thought, learning and language, and lashes with a holy scorn the crying sin of the time, hypocrisy turned inside out. Read George Herbert's church-porch or his Cornaro, or his 'brother' Nicolas Ferrar's Lessius,[21] and you will allow that Cambridge tradition is no friend to excess. From the gentile world, the Seven Sages, Socrates, Plato, Epicurus himself, Seneca, Epictetus, Porphyry, the same lessons are learnt. *These all, not having the law, did by nature the things contained in the law.*

Are these teachings obsolete now? Is there no need for a homily of fasting or of gluttony, for a commination service or Lenten collects?

Let us summon witnesses: and first the ecclesiastical head of the Romish church in England, who in his pastoral for this season[22] denounces: 'drunkenness which inebriates man and woman, and is spreading even to children ; crime and vice of every kind, which spring from the brain maddened with drink ; luxury and excess, gluttony and self-indulgence, hardness of heart and selfishness both refined and rude: and these things are pervading our society in all its classes.'

Or would you hear a lay voice ? The venerable president of a society for the reform of diet, writes[23] : 'The money spent on dinners (and other meals) is lamentable; I almost say, disgraceful. Most certainly the public health is immensely lowered and life shortened by prevalent excess of eating, unawares accepted as normal.'

Are these heated partisans ? Turn to the cold science of Germany. Prof. Friedländer[24] will teach you that the luxury of imperial Rome was as nothing to that of imperial Berlin. What then shall we say of London? Look to other facts patent to the world. Our regius professor of physic watched the death-rate of the cotton districts during the cotton famine : it was far lower than the average. Wealth in Manchester is not health. The prisoners in our gaols are the healthiest class among us. Freedom to Englishmen is not health: to make them healthy you must enslave them. When the leader of the Brahmo Somaj, Keshub Chunder Sen, took leave of our country, he could not forbear lifting his voice against our groaning tables.[25] Our nurses in Turkey and at the Cape were amazed at the readiness with which wounds healed in their hospitals. And why? Asiatic and Zulu bodies were not inflamed by high feeding. 'A multitude of dishes,' says Seneca,[26] ' is the cause of the multitude of diseases.' 'The heroic age,' says Celsus, after Plato,[27] 'enjoyed good health, because it was a stranger to sloth and luxury.' 'This heavy-headed revel, east and

west, makes us traduced and tax'd of other
nations.'

To whom shall we go for health? To the
physicians? Yes, to the greatest of them. Hip-
pocrates[28] will tell you: 'It is easier to be filled
with drink than with meat.' In other words:
Over-eating is more dangerous, clogs the system
more, than over-drinking. Galen, Hoffmann,
Hufeland, Cheyne,[29] and many another glory of
the healing art, declare that temperance is not
only true luxury, but the only physic. Man's
body was not made to be the battlefield of
drugs. But for many years other doctrines have
prevailed: it is nothing that prophets of despair
preach on platforms and in books[30] that continence is impossible for man; it is nothing
that charlatans warrant sin without sorrow, de-
bauchery without its sting: but it is surely
strange that quiet, family doctors should prescribe drunkenness to temperate ladies.[31] If my
witnesses are to this day strictly sober, they
owe no thanks to the faculty.

So we might interrogate many oracles and
hear ambiguous responses. If bards have hymned
in fine frenzy the praises of Bacchus and Venus,
there is also a staid and a heavenly Muse, which
weds, like our Johnian poet,[32] 'plain living and

high thinking'; which cries with the Country
Parson,[33] 'Slight those who say amidst their
sickly health, Thou livst by rule. What doth
not so but man? Houses are built by rule
and commonwealths. Entice the trusty sun,
if that you can, From his ecliptic line; beckon
the sky: Who lives by rule then, keeps good
company.'

Only when we consult sages and saints, do
we hear a clear and harmonious voice: *Let
us who are of the day, be sober.* Wise men
of the East and West, Stoic and Epicurean,
Fathers and Reformers, Fisher and Luther,
William Law and John Wesley,[34] Thomas Ar-
nold and John Keble, are all at one in the
doctrine and the practice of strict temperance, so-
berness and chastity, as binding on all, possible
to all. *Be ye holy, for I am holy*, is no cruel
irony, but a promise and means of grace. *His
strength is made perfect in our weakness.*

Our church, like the Old Catholic churches
of Germany[35] and Switzerland, issues no command to fast, no table of prohibited meats. As
in confession to the priest, so in fasting, both
churches leave private Christians free to consult
their own reason and conscience, and individual
needs. To us *touch not, taste not, handle not,*

must be the dictate of ripe resolution and free surrender. 'Fasting,' says the homily, 'of itself is a thing merely indifferent, but is made better or worse by the end that it serveth unto.' How shall we use our freedom in this matter? In fashionable shops I have seen an ingenious toy, called a discipline. It is a scourge for the use of penitents, copied from mediaeval models. Shall we class the Lent fast with the will-worship of these new flagellants, or has it any lesson for men of reason here and now?

*Plures gula quam gladius.*[36] The sword slays its thousands, gluttony its ten thousands. So deadly is this mysterious sin, seldom named to ears polite. Shall we be among its victims? Epicurus recommended, Seneca[37] and other wealthy Romans practised, the occasional adoption by the rich of the poor man's life, the coarse fare, the hard bed, the mean clothing. Might not we each Lent deny ourselves some selfish luxury, and buy with the saving the luxury of doing good? 'Anglicans never inculcate the duty of almsgiving'; so I heard Father Faber assert in the Oratorian church. I fear that among us there may be some colour for the sarcasm: men may come and go and be often taxed for the pleasures of the rich, but never once exhorted 'to help the humble store' or 'mend the dwellings, of the poor.'

Lent also bids us reverence the temples of the Holy Ghost, so fearfully and wonderfully made—our own bodies. Socrates[38] bids us avoid what tempts us to eat when not hungry or drink when not thirsty. It is a physiological law, which any man can verify for himself: the natural supply of natural wants, as bread and water, soon contents us;[39] we have enough, which is better than a feast; whereas artificial diet, intoxicant, narcotic, opiate, creates an artificial craving. 'The drunken man is always dry.' If you are in the habit of taking stimulants of any kind, for your own freedom's sake try the experiment of abstinence this Lent; and if the abstinence is difficult, make it perpetual. If there is one point on which medical science is agreed,[40] it is that to some constitutions alcohol in every form is poison. Dr. Johnson frankly confessed[41] 'I cannot take a little.' If any here has been overcome with drink, let him take the warning while there is time; and he may learn to thank God that we have still a Lent.

Lent has a word to say respecting our gifts. Sallust[42] remarks: 'Long since we have lost the true names of things; we give away other men's goods and call it generosity.' Be just before you are generous. Most students are sent to college only by a parent's self-denial. Repay the sacrifice in kind by a Lenten frugality. It is the tradition of the place. The very stones of our buildings, the very stipends of fellows and scholars, tell of Johnian self-conquest and public spirit in the past. The new court arose because fifty years ago our fellows were content with one-third of the present dividends.

One advance has certainly been made in this matter of gifts. Many a Cambridge shopboy owes his ruin to orders on the butteries, given in thoughtless good nature. I hope that this cruel kindness is now all but obsolete. Is a like gift to friends, to be consumed in dangerous excess in a limited time, altogether unknown?

Contrast with such mock friendship an example of true generosity. Dr. Paley[43] is the speaker: 'I spent the first two years of my undergraduateship happily but unprofitably. I was constantly in society where we were not immoral, but idle and rather expensive. At the commencement of my third year however, after having left the usual party at rather a late hour in the evening, I was awakened at five in the morning by one of my companions, who stood at my bedside and said: Paley, I have been thinking what a fool you are. I could do nothing probably, were I to try, and can afford the life I lead: you could do everything, and cannot afford it.' Paley took the rebuke in good part, rose from that time forward at five, read all day long with intervals for chapel and hall, and graduated first as senior wrangler, then as a teacher, philanthropist and author of the highest class.

O that we could follow in these matters of daily life that maxim of the African church:[44] 'Our Saviour Christ called Himself not custom but truth.' 'Evil is wrought by want of thought, As well as want of heart.' Students here and elsewhere entertain their friends sumptuously, never dreaming that they are giving what is not theirs to those who do not want. Such mercy curseth him that gives and him that takes. Meanwhile there is real want all around. *I was an hungered and ye gave Me no meat.*

Some years ago a wealthy undergraduate here spent £70 on a dinner. The fault was not his, for we had never shewn him Addenbrooke's hospital or the industrial school. He knew not that poverty was our badge from the

first, not that *ambitiosa paupertas*[45] which apes
the follies of the rich, but the hardy nurse of
Aschams[46] and Chekes and Cecils. In riper
years, when ashamed of his wilful waste, he
could only hope that it had not brought to
woeful want some poorer friend.

As men count excess, it cannot be said that
our college table, even as it is now, provokes to
it. Last year a German, present at a college
feast, was surprised that, amidst so much abundance, the guests took so little. Waiving the
question whether intemperance does not begin
far short of the line drawn by the fashion and
science of the day, let me address a concluding
caution to our *imperia in imperio;* for a sober
college may be ruined by riotous clubs. Our
chief club was founded some fifty years ago
by men who knew the temptations of good-
fellowship: they therefore named it by the
name of the lady Margaret, as a talisman
against luxury,[47] for to the athlete luxury is
suicide. With the foundress join the memory
of the hero bishop of the isles, and athletics
may be a help, not a hindrance,[48] to the only
end of life which Margaret of Richmond or
George Selwyn[49] prized : that we may offer and
present ourselves, our souls and bodies, dis-
ciplined to bear and to forbear, to be a reasonable, holy and lively sacrifice unto God.

# NOTES TO INTRODUCTION.

[1] p. 6. Roger Ascham. Sein Leben und seine Werke von Dr. A. Katterfeld. Strassburg, Trübner, 1879, p. 25 : 'Welch ein Unterschied zwischen diesem Bilde und dem reichen behäbigen Luxus, der dem Besucher heute überall in Cambridge entgegentritt !'

[2] p. 8. Keshub Chunder Sen in the Dietetic Reformer, 1870, p. 108. A missionary told me that an Indian said to him, 'You are not a Christian.' 'Yes, I am.' 'No, you can't be ; you don't drink brandy ; you don't eat beef.'

[3] p. 8. It is a good sign of the times that the medical publisher, Baillière of Paris, has reprinted Cornaro and Lessius with 'le régime de Pythagore d'après le docteur Cocchi.' Cocchi's book is brought down to the present time, and the temperance reform is frankly and cordially welcomed. *Les principes*, says the preface, *que Cocchi, Cornaro et Lessius ont exposés et développés sont éternellement vrais. La meilleure preuve nous en est donnée par les sociétés de tempérance qui fonctionnent en Amérique et en Angleterre depuis une cinquantaine d'années, et par les sociétés de végétariens qui se fondent en Allemagne et en Suisse ; elles témoignent d'un mouvement très-intéressant et très-répandu de l'opinion publique dans le sens de la vie sobre et rendent de grands services aux populations laborieuses, en même temps qu' elles servent la cause de la civilisation générale.*

[4] p. 8. An undergraduate had stated that he had been a vegetarian for three years, and long hesitated about entering the university, fearing that plain living was tabooed among us.

[5] p. 9. Printed on the reverse of the half-title.

p. 9, L 13 from bottom, on the words *endowed at the expense of the colleges*, add :     '

It is remarkable that no professor of Health (but of Pathology) has been demanded by the faculty.   Compare the complaint of a surgeon, Mr. J. J. Pope, read at the Society of Arts 5 March 1879 (House and Home, 12 April 1879, p. 139): " In our system of medical education we have all along omitted the subject of hygiene.  Our student is taught to view and investigate the body in endless varieties of diseased conditions, but has been left in complete ignorance of what it is in a state of health ; and although this fearful oversight has been recognised at last, in one or two of our principal colleges, and copying the example of the Army Boards, chairs of public health have been established, and courses of instruction afforded by competent professors, yet no examination on the subject is held to be necessary, and the matter still occupies, as it were, an extraneous and ornamental position.   What is required, and what must be done, is to edu-cate the masses in these things, when they will insist on the beneficial adoption of sound sanitary principles, they will clamour for the enforcement of the regulations now almost in abeyance."

Dr. Nichols (Herald of Health, 1 May 1880, p. 350) : " Our medical customs make it the interest of doctors, surgeons and chemists, that people should be ill rather than well.  Doctors are not paid to teach people the laws of health and to prevent disease, and, as a rule, they do simply what they are paid for doing."

In the Vegetarian Messenger X (1860) under the title 'homœo-pathy and something more' is reprinted from the British Journal of Homœopathy an article by Dr. E. Ackworth, which tells many home truths :

" Health is something more than recovery from sickness. Better than any cure is not requiring one.   Better than *getting* well is *being* well.   Yet how much more cure is thought of than prevention. . . . Health is only to be had by observing its con-ditions, of which taking physic is not one.   Taking physic *implies* disease. No one that is well requires to take it. . . . Health is but obedience to the laws of physiology, in other words, to the laws of GOD. Disease is the penalty attached to their transgression. 'When transgressed, the penalty cannot be evaded. . . .

"We treat effects, but tolerate their causes. . . . The misfortune is, the causes of disease seem such a difficult subject of inquiry. But would they seem such if common sense (leaving science altogether out of view) were a little more brought to bear thereon, and we thought a little more harm than we think of running counter to the laws of our own nature? Surely, in such cases for instance, it hardly could be mooted (and that in medical journals too), whether smoking tobacco were injurious or not. . . . Surely the giddiness, the vomiting, the fainting— the first effects of a poison, in short—are as much nature's protest against its use as anything short of death well can be. . . And that, little by little, we *can* grow accustomed to that which, in health, was first found to be injurious, is an argument one would think, in medicine as in morals, *against* growing so accustomed. . . .

"Medicine may dally with this or that symptom and relieve an ache or pain for a while; but in a large sense restoration to health implies the restoration of unperverted instincts. . . .

"Now it is with man as with the lower animals. In his natural, not his savage, state, he would know what 'to eat, to drink, and to avoid.' But his boasted reason sophisticates his instincts. . . . He falls into habits that become his second nature, and follows his second nature as if it were his first. Now, just as he does so—just as he recedes from the rule of his first, and follows that of his second nature—does disease, with all its sufferings, ensue. The first bids him eat to satisfy his hunger—and a sound digestion waits on his plain and proper fare. The second bids him eat less to satisfy his hunger than to gratify factitious wants and the wretched cravings of a depraved appetite—and dyspepsia is a necessary and natural result. . . .

"Every day of our lives, in the words of our prayer-book, we are 'leaving undone those things which we ought to do, and doing those which we ought not to do, and there is no health in us.' . . . . How is this to be corrected by a dose or two of physic? . . . . Pathology is but physiology neglected. . . .

"Until we set ourselves seriously to work to reform the evil habits of society, and bring the modes of civilised life into strict obedience to hygienic laws, instead of those of idiotic fashion, we may just as well 'throw physic to the dogs,' for any great good the world may hope to gain from it. . . .

"On this subject of weakness there is a vulgar error—a vulgar and most pernicious error—into which physicians, and not alone the public, are only too apt to fall. And this error lies in supposing weakness—no matter on what cause that weakness may depend—is to be overcome by taking large quantities of food, and chiefly food of a stimulating kind. No notion than this can possibly be falser—more opposed to all that physiology would teach us. Food is not the sole element of strength, and may be, and very often is, an element of weakness. We shall find that, as far as food is concerned, strength is a thing that not more depends on what is taken into the body than what is carried out of it. . . . A *glut* has much the same effect on the animal system as it has on the commercial. The excess, in either case, produces loss. . . .

"Were man to live as he ought to live, he would die of old age (that rarest of complaints) with scarcely an ache or pain. But he does not live as he ought to live, and seems whilst living as viciously as may be (not morally perhaps, but physiologically speaking) to think that medicine is a form of absolution that enables him to do so without suffering for his sins. And whilst he is allowed to repose in this belief, as we fear he too often is by our profession, what wonder if the charlatan find favour in his sight, who is ever ready with his form of absolution—some saving globule, draught, or pill—rather than the upright and well-informed physician who tells the patient the honest truth, and lets him know how impotent is medicine when given to supersede amended life? Whilst human nature is what it is, it will shew itself in medical as well as higher matters, by a preference for those who respond to its cry of ' Prophesy unto us smooth things '—' Let us eat and drink—no matter what or how —and be as if we did not.' The public—or at least a large portion of it—would like ' to lay the flattering unction to its soul ' that health is to be had without observing its conditions. It would like *to be well*, and yet *do ill*—to live, but not to feel, amiss. It can't. Those whose god is their belly must pay for their idolatry—those who live to eat cannot enjoy the same immunity from suffering as those who eat to live. . . . And if we would do our duty to the public, we should let it know more than it knows at present of the way, or rather the various ways, in which it violates the laws of health, and how impossible it is for medicine to enable it to do so with impunity."

[6] p. 11.   A happy *paronomasia* μᾶλλον ἔμπειρον εἶναι τῶν εὐστομάχων ἢ τῶν εὐστόμων.

[7] p. 18.   VI 480 *sunt quae tortoribus annua praestant.*

[8] p. 28.   See ' Reasons against vivisection ' by Prof. Newman in T. L. Nichols' Herald of Health 1877, p. 209, or Dietetic Reformer 1876, p. 185.

[9] p. 28.   See a letter (30 July 1876) addressed by Dr. Haughton to Mr. Sclater-Booth, *e.g.* (T. L. Nichols, Herald of Health, 1876, p. 54) *There is no country where such laws exist that does not contain educated medical men who regard their operation as highly injurious to the health of the people. . . . The same favour which now is extended to vaccination by the medical corporations was formerly given, and quite as unreservedly, to the practice of inoculation with small-pox matter (now declared penal), and they continued to support this monstrosity for nearly a hundred years.* See ibid. p. 87, where Dr. Nichols says : *There is no doubt that many diseases may be, and are, propagated by vaccination. No skill or care of the vaccinator can prevent it. There is no doubt that many children are killed by vaccination. There is, perhaps, as good reason for making vaccination a penal offence, as there was for passing an act of parliament against inoculation.* ibid. p. 107

seven Keighley poor law guardians in York castle. Ulcers caused by vaccination. ibid. p. 179 Dr. Thomas Skinner writes of two children : *They were victims of the law, and died from vaccination. . . . Why should anyone in a Christian country be ashamed to own and tell the truth when called upon? and especially the law and medicine—our protectors? . . . . A young lady, aged fifteen, a charming girl, and dearly beloved by all who knew her, was vaccinated by myself, with vaccine lymph as pure as any. . . . Eleven days after the operation this blooming girl, who was in the best of health, died from nothing more or less than . . . . re-vaccination. . . . I brought the case before the Medical Institution in this town. The meeting of members owned that the death was due entirely to vaccination, but they expressed a desire that it might not be made public.* Mr. W. G. Ward (The Vaccination Inquirer Nov. 1879) in defending his own child, cited from the neighbourhood examples of 'cancerous syphilis' and other diseases due to vaccination. *e.g., syphilis has undermined the nose, and eaten a passage under the left eye.* A father in Cambridge shewed me his son, the picture of health. ' Yes, but it has cost me £50 to save him from cow-pox.'

At the Dialectical Society, 7 April 1880 (Herald of Health, 1 May, p. 345) "Dr. Collins, for twenty years a public vaccinating officer, gave his testimony to the effects of vaccination which had compelled him to abandon and oppose the practice at a pecuniary sacrifice of £500 a year. He proved the inefficacy of vaccine by putting it in one arm and small-pox in the other, and finding that both diseases went on together in the same subject. He found also, that while some persons could not be vaccinated nor infected with small-pox, others would take either infection again and again."

The Vaccination Inquirer Oct. 1876, p. 97 contains a paper by Mr. Henry Pitman ; 'How I became an Anti-Vaccinator :'

*My wife and I allowed our first-born children to be vaccinated as a matter of course. Like most young couples, we had not studied the question physiologically or politically, and never dreamt of questioning its rightfulness. It was considered to be the doctor's business, not ours. . We have since learned that it is perilous for people to submit their bodies to the doctors or their souls to the priests.*

*The vaccination of our second daughter was followed by a shocking abscess under one of her eyes, which threatened to destroy her sight. This led me to think, to read, and inquire into vaccination. I discovered that abscesses, ulcers, and erysipelas, often followed the infusion into the blood of the corrupt matter miscalled "lymph." I got a second-hand copy of Dr. Jenner's "Inquiry" (which cost me half-a-guinea), and the description he gave of the effects of vaccination convinced me that it was a much worse infliction than small-pox, which is not a disease to which infants are particularly liable; in fact, more children die from burns and scalds than from small-pox. . . .*

*Notice was served to have our youngest child vaccinated. I told the officer that no more of our children should be "blood-poisoned," whatever the consequences might be to myself. . . . "I suppose you want to be a martyr," said the Stipendiary. "I want to protect my child from disease," was my reply. The full penalty was inflicted; 20s. and 10s. costs. I am sorry to say I paid the fine.*

How many medical men, how many of those by whose votes the Vaccination Acts were passed, have taken half the pains to examine into the merits of the question that this persecuted father has done? Supposing it were true that a million children were saved by poisoning this one healthy child, what barbarous creed would sanction such vicarious sacrifice? *Tantum relligio potuit suadere malorum.*

Mr. Pitman (ibid. Nov. p. 114) continues : *Being of a combative nature, I attacked the law by publishing the* Anti-Vaccinator, *with this motto from Dr. Garth Wilkinson's masterly writings on the subject : " Other wars are towards death, but in this crusade the war is against death." . . . . Of course I became a "marked man," and was summoned again ; and, declining to pay the fine, was sentenced to fourteen days' imprisonment. How I fared is told in my Prison Thoughts on Vaccination, which I shall be*

5

*happy to send freely to any applicant* (Mr. Pitman's address is 41 John Dalton Street, Manchester).

Dr. Edward Haughton (ibid. 100-1), speaking on oath in court, enumerates syphilis, scrofula, leprosy, eczema, erysipelas, as transmitted by vaccination.

[10] p. 29. Sir Thomas Watson says in the Nineteenth Century: *I can readily sympathise with, and even applaud, a father who, with the presumed dread and misgiving in his own mind, is willing to submit to multiplied judicial penalties rather than expose his child to the risk of an infection so ghastly.*

[11] p. 34. A tradesman here told me that former voluptuous customers lamented the tastes from which they had a difficulty in weaning themselves as country curates.

[12] p. 34. Once only have I taken medical advice. I was possessed by thoughts which gave me no rest night or day; meal-time passed by unheeded and I craved no sleep. A justly honoured physician prescribed whisky as a 'nightcap.' It had no effect whatever. A clerical friend put a kettle on the fire, ran a gutta-percha tube under a chair, on which I sat in blankets. This home-made vapour-bath brought sleep at once. Twice I myself exorcised the same fiend; once by going into the country and attacking an entirely new study; once by walking round the college garden bareheaded, looking up at the birds and down at the flowers, until my head cooled down and the tyrant thoughts fled. Many of the members of the Vegetarian Society have passed through the ordeal of a 'generous' diet 'ordered' by the faculty.

Mrs. Nichols (Herald of Health 1880, p. 342) testifies: "We often have patients who have been counselled to eat flesh five or six times in a day, and to take spirits or malt liquors beside; and this when the poor stomach needed rest, and was hardly able properly to digest a few ounces. To unlearn the lessons of ignorant teachers and to undo the mischief done by ignorant medical advisers is a hard, sad piece of work, but it must be done, and in no case is it more imperative than when flesh-meat and spirits have been prescribed as health-giving."

[13] p. 39. From his childhood Walker was sickly. Learning from Cicero that health is very much in our own power, he exclaimed 'I will be well.' In his famous book he explains the art and also gives many valuable hints to the beneficent.

[14] p. 40. Catil. 2 § 5 *imperium facile eis artibus retinetur, quibus initio partum est.*

---

## NOTES TO SERMON.

[1] p. 45. *two discourses by Edw. Brome.* See the title in Tho. Baker's Hist. of St. John's coll. (1869) p. 1000 : the dedication p. 1001.

[2] p. 45. *then near his end.* He died 27 March 1711 (ibid. 995).

[3] p. 45. *Ambrose Bonwicke* (ibid. 998-9. I reprinted his life in 1870, Cambridge, Deighton, Bell and co.)

[4] p. 45. *Gunning's discourse on the Lent fast.* A presentation copy is in the college library, and it was reprinted in the library of Anglo-Catholic theology, Oxford, 1845. It is founded on a sermon preached in Lent before Chas. II. to whom it is dedicated.

[5] p. 46. *our historian,* pp. 239, 660.

[6] p. 46. *Gower tells us* (ibid. p. 655).

[7] p. 46. *Jenkin, in his Latin character of Gunning* (ibid. p. 656).

[8] p. 47. *the presbyterian demand,* Gunning, ch. 7. Cardwell, history of conferences, p. 306.

[9] p. 47. *fasts on Dec. 25.* See Hen. Vaughan's poems (1847) p. 213, 'the nativity, written in the year 1656.' In 1643 (Baillie's letters II 120) 'all of us stoutlie had preached against their Christmass . . . both houses [of parliament] did profane that holy

day, by sitting on it, to our joy, and some of the Assemblie's shame.' Wordsworth, Christ. Inst. iv 488, 551-2, 660.

¹⁰ p. 47. *Cromwell's soldiers on Christmas day,* 1657, *made prisoners of Gunning's congregation.* Evelyn's diary under that date; Kennett's compl. hist. 223.

¹¹ p. 47. *Chrysostom* de paenit. hom. 5, c. 4 (II 371, Gaume, cited by Gunning, 119-159; there is much more to the point in Chrys. ib. c. 4 and 5). Elsewhere he says (ad pop. Ant. hom. 3, II 47*b*, ed. Gaume), φάρμακόν ἐστιν ἡ νῆστεια. in genes. c. I, hom. I (iv 5*a*), 'for if the sons of the physicians, when about to give medicines to those who wish to purge away the putrid and corrupt humour, bid them abstain from this bodily food, that it may not become a hindrance to the virtue of the medicine.' See the Benedictine indexes to the fathers under 'ieiunium.' Cf. Hippocr. aph. II 4 and 17 ('when more food than is proper has been taken, it occasions disease'; where see Adams II 708). Cels. II 16 *ubi ad cibum ventum est, numquam utilis est nimia satietas; saepe inutilis nimia abstinentia; si qua intemperantia subest, tutior est in potione, quam in esca.* II 16, *neque ulla res magis adiuvat laborantem, quam tempestiva abstinentia. intemperantes homines apud nos ipsi cibi tempora curantibus dant.*

¹² p. 48. *Augustus* (Suet. 76). *Vespasian* (Suet. 20, one day in every month).

¹³ p. 48. *another use of fasting.* The connexion between excess in eating and drinking on the one hand and lust on the other is attested by universal experience. Thus Socrates asks Aristippus (Xen. mem. II 1 § 16) : 'How do masters deal with such slaves? ἆρ' οὐ τὴν μὲν λαγνείαν αὐτῶν τῷ λιμῷ σωφρονίζουσι;' The line of Terence (eun. 732, where see Lindenbrog and Ruhnken), *sine Cerere et Libero friget Venus,* became proverbial. Erasmus (adagia, Francof. 1656, 458-9) illustrates it from the Greek drama, *e.g.* from the fragment in Athen. 28*e* :

ἐν πλησμονῇ γὰρ Κύπρις, ἐν δὲ τοῖς κακῶς
πράττουσιν οὐκ ἔνεστιν 'Αφροδίτη βροτοῖς.

See also Eur. Bacch. 773, Tert. adv. Marc. II 18, Clem. Al.

paed. II 2 §§ 20 21, Apul. m. II II *Veneris hortator et armiger Liber*, where see Price (vol. III, p. 175*b* Oud.)  Ov. a. a. III 762. Anthol. Lat. 710 3, 4 R :

> *ardenti Baccho succenditur ignis Amoris,*
> *nam sunt unanimi Bacchus Amorque dei.*

¹⁴ p. 48. *Leo the Great* de ieiunio decimi mensis II (serm. 13, al. 12). With the next citation cf. Clem. Al. paed. II 1 § 7 τὴν αὐτάρκειαν, ἣ δὴ ἐφεστῶσα τῇ τροφῇ, δικαίᾳ ποσότητι μεμετρημένη σωτηρίως τὸ σῶμα διοικοῦσα καὶ τοῖς πλησίον ἀπένειμέ τι ἐξ αὐτῆς.

¹⁵ p. 48. *Basil's words* cited by Gunning, p. 162=218.

¹⁶ p. 48. *Gunning*, pp. 168-174=227-235.

¹⁷ p. 49. *Dr. Wm. Lambe* of Herefordshire, educated at Hereford school under Mr. Rudd, admitted pensioner of St. John's college 14 March 1782, under Messrs. Pearce and Kipling ; admitted scholar on the duchess of Somerset's foundation 1782 ; B.A. (4th wrangler) 1786, M.A. 1789, M.D. 1802 ; admitted fellow of St. John's, on lady Margaret's foundation, 11 March 1788 ; his successor was elected 7 April 1794. He married at Warwick Miss Welsh, daughter of Captain Welsh (*Cambr. Chron.* 8 March, 1794). See Memoirs of Physicians (3rd ed.) 1822 ; Biogr. Dict. of Living Authors, 1816 ; Public Characters (1823) II 542 ; Munk's Roll of the College of Physicians, III² 17-18 ; esp. the Life of William Lambe, M.D. by E. Hare, C.S.I. inspector general of hospitals. Reprinted from *The Dietetic Reformer.* London, F. Pitman ; Manchester, 56 Peter street. 1873, 8vo. Several of Lambe's scarce books are in the University Library.

¹⁸ p. 49. Dated 21 Aug. 1563, p. 265, Parker Soc.

¹⁹ p. 49. 'The house of feasting' and 'apples of Sodom' are in his 'twenty-five' sermons, nos. 15, 16 and nos. 19-21. In the 'Life of Christ' see part I sect. 8 disc. 4, 'of mortification and corporal austerities,' where he says (§ 2), *Mortification is the one half of Christianity ; it is a dying to the world.* Part 2, sect. 12,

disc. 13, 'of the manner of fasting.' See also 'Holy Living' ch. 4 sect. 5, 'of fasting.'

[20] p. 49. In the Cambridge ed. 1859, IV 142-241, *e.g.* p. 173 *The declension of piety is not perhaps more to be ascribed to any other cause than to this, that men who approve goodness in their hearts are so backward to shew it in their practice.* p. 177 *Exemplary and edifying conversation is a debt which we owe to the world, a good office imposed on us by the laws of common humanity.* p. 181 *A good conversation before men is a part of that due respect which we owe to them. There is a regard and a kind of reverence to be had toward every man.* p. 190 *We shall highly oblige those whom by our good endeavour or example we shall convert to righteousness, or reclaim from iniquity, or shall anywise stop in their career to ruin.* p. 214 *It is indeed a sad thing . . . that it should become desirable that hypocrites might abound in the world, lest religion both in truth and show should be discarded.*

[21] p. 50. I have (a copy presented by Barnabas Oley of Clare, the friend of Ferrar) 'HYGIASTICON : | or, | The right course of | preserving Life and | Health unto extream | old Age : | Together with soundnesse | and integritie of the | Senses, Judgement, | and Memorie. | ¶ Written in Latine by | *Leonard Lessius,* | And now done into | English. | *The second Edition.* | Printed by the Printers | to the Universitie of | *Cambridge.* 1634.' sm. 12mo. The book contains also that famous piece of Cornaro's 'Treatise of Temperance, translated by Master George Herbert'; and 'a discourse translated out of Italian.' See Two lives of Nic. Ferrar, Cambr. 1855, ind. s. v. *Lessius.* Backer, Bibliothèque des ecrivains de la compagnie de Jésus, under *Lessius,* records many editions of this excellent tract down to our own day. Baillière in Paris has just issued a French version of Cornaro and Lessius. The burden of both books is, in Crashaw's words :

> *That which makes us have no need*
> *Of physic, that's physic indeed.*

[22] p. 50. Canon Quinlivan kindly procured a copy of this document for me from cardinal Manning. The substance of it

may be seen in the Tablet for February, 1880. A pastoral letter to the clergy and laity of the diocese of Westminster, for Quinquagesima Sunday. By Henry Edward, cardinal archbishop of Westminster. London, 1880. There are several remarkable passages beside that in the text. *Seven millions of armed men are watching each others' movements with mutual suspicion and fear. Can such a gathering of storm-clouds disperse without an outbreak which will lay waste half the christian world?* . . . *In no country and in no age has the world as yet ever seen such commercial activity and prosperity as that of England. But in the midst of immeasurable wealth is a want which the poorest country of Europe scarcely knows. We have in the midst of us not poverty alone, which is an honorable state when it is honest and inevitable, but also pauperism, which is the corruption of poverty and the debasement of the poor.* See Dr. Nichols' 'How to live on sixpence a day,' p. 23 *There are, to take a noted and living example, few harder working men in England than archbishop Manning, a man full of cares and labours, yet I am assured by those who have had the most intimate personal relations with him, that Mr. Disraeli, in* 'Lothair,' *has not in the least exaggerated his habitual abstinence, and that his ordinary meal in public or private is a biscuit or a bit of bread and a glass of water.* The distinction between poverty and pauperism (*paupertas* and *egestas*) is admirably drawn by Thos. Walker, The Original no. 13 (II 52-55, ed. Bl. Jerrold, Lond. Grant & Co. 1874).

[23] p. 50. *Emeritus* Prof. F. W. Newman, of Weston S. M. president of the Vegetarian Society, in a late letter to me.

[24] p. 51. Darstellungen aus der Sittengeschichte Roms in der Zeit von August bis zum Ausgang der Antonine. Von Ludwig Friedländer. Leipzig, 1871. A new edition of this third volume is announced. See pp. 1-104, 'der Luxus.'

[25] p. 51. This address is printed in the Dietetic Reformer of 1870, pp. 108-110. Cf. 1871, pp. 34-39. Compare Strabo's complaint (VII 3 7 fin. p. 301) of the corruption of barbarians by Greek luxury.

[26] p. 51. Sen. ep. 95 18 *simplex erat ex causa simplici vale-*

*tudo: multos morbos multa fercula fecerunt.* ibid. 2 4 *fastidientis stomachi est multa degustare, quae ubi varia sunt et diversa, inquinant, non alunt.* Aristot. problem. I 15 ἡ ποικίλη τροφὴ νοσώδης (ταραχώδης γὰρ καὶ οὐ μία πέψις). Hor. s. II 2 70-79 (Conington's version : cf. the whole satire) :

> *now listen for a space, while I declare*
> *the good results that spring from frugal fare.*
> *imprimis, health : for 'tis not hard to see*
> *how various meats are like to disagree,*
> *if you remember with how light a weight*
> *your last plain meal upon your stomach sate.*
> *now, when you've taken toll of every dish,*
> *have mingled roast with boiled and fowl with fish,*
> *the mass of dainties, turbulent and crude,*
> *engenders bile, and stirs intestine feud.*
> *observe your guests, how ghastly pale their looks,*
> *when they've discussed some mystery of your cook's :*
> *ay, and the body, clogged with the excess*
> *of yesterday, drags down the mind no less,*
> *and fastens to the ground in living death*
> *that fiery particle of heaven's own breath.*

Celsus (III 6) denies the assertion of Asclepiades, that a varied diet is more easy of digestion. Plut. VII sap. conv. 4 fin. (II 150) 'it rather diminishes than increases one's expense, to entertain wise and good men, for it takes away superfluous dainties, and foreign ointments and cates, and broachings of costly wines.' The question whether varied or a simple food is more easy of digestion is treated at length by Plut. qu. conv. IV 1 and Macrob. VII 4. Cf. Clem. Al. paed. II 1 § 2.

²⁷ p. 51. Cels. praef. of the Homeric age *verique simile est inter nulla auxilia adversae valetudinis, plerumque tamen eam bonam contigisse ob bonos mores, quos neque desidia neque luxuria vitiarant. siquidem haec duo corpora prius in Graecia, deinde apud nos, afflixerunt.* Plato republ. 405c-408e, legg. 720, 857 (these passages against mere empirics).

²⁸ p. 52. Hippocr. aphor. II n. 10 ῥᾷον πληροῦσθαι ποτοῦ ἢ

σιτίων. [Cels. I 2 p. 15 l. 14, Daremberg.] Cf. 9 τὰ μὴ κα-
θαρὰ τῶν σωμάτων ὁκόσῳ ἂν θρέψῃς μᾶλλον μᾶλλον βλάψεις. 17
ὅκου ἂν τροφὴ πλείων παρὰ φύσιν ἐσέλθῃ, νοῦσον ποιέει. II 51
πᾶν τὸ πολὺ τῇ φύσι πολέμιον. epidem. VI 4 18 ἄσκησις ὑγιείης,
ἀκορίη τροφῆς, ἀοκνίη πόνων. [de affectionibus] 50 danger of
excess even in wholesome meat and drink. de flatibus 7. On
the use of abstinence cf. Celsus pr. p. 6 l. 13-21 Daremberg.

²⁹ p. 52. It is much to be wished that some one would draw
up a code of health from GALEN. See directions for abstinence
(de sanitate tuenda III 13, vol. VI 231 K, IV 4 pp. 247, 259, 260);
cures effected by an obedience to Galen's rules of health (ib. V I
p. 308); intemperance and ignorance the causes of disease (ib.
ad fin. pp. 311-12); the mean (τὸ σύμμετρον) to be observed in
eating and drinking (ib. 2, p. 313); exercise before dinner (ib. 3,
p. 319 καλῶς εἴρηται τοὺς πόνους τῶν σιτίων ἡγεῖσθαι); the physician Antiochus, aet. 80, his diet of bread and honey, fish and
fowl (ib. 4 pp. 332-3); Telephus the grammarian lived to near
100, his diet was porridge with honey, vegetables, fish or fowl,
bread dipt in wine (pp. 333-4); obstructions from various kinds
of food (ib. 6 pp. 339-40); the old man who takes care of himself will not need medicine, but may cure himself by a spare
diet (p. 341); fruits and vegetables as purges (ib. 9 p. 353);
should we eat once a day or twice? (VI 7 p. 410); Galen's rule,
when his visits to patients or business made him postpone his
bath to 4 or 5 p.m. to take breakfast at 10 or 11 of bread only;
others took also dates, olives, honey or salt (p. 412); dangers of
intemperance at feasts (p. 415); the art of health promises unbroken health to such as obey it; to the disobedient it is as
though it did not exist (ib. 8, p. 415).

FRIEDRICH HOFFMANN (1660-1742) *opera omnia*, suppl.
II (1) Gen. 1753, fol. pp. 484-492 *diss.* (1728) *de medicis mor-
borum causa.* Vol. V (1748) 328-333 *de inedia magnorum mor-
borum remedio.* Here he cites the 'elegant medical adage,'
*modicus cibi, medicus sibi;* also the long lives of the Essenes
(Josephus b. J. II 7) and of Cornaro, due to temperance. Re-
buking the luxury of his age, he cites ecclus. 38 33 and Sen.

ep. 95 § 23 *innumerabiles esse morbos non miraberis : coquos numera.* §§ 15-18, he complains of his brethren : *plurimi practicorum gulae mancipia sunt, aegrorum luxui adulantur, ipsisque aegrotantibus persuadent, ipsos neglecta diaeta sanitatem recuperare posse solis pharmacis atque medicinis, quas speciosis titulis ornare et ad caelum usque ipsarum virtutes laudibus extollere solent.* He commends the ancient physicians for their reliance on the diet cure ; quotes Celsus III 21 (p. 106 Daremberg) of dropsy : *facilius in servis, quam in liberis, tollitur : quia, cum desideret famem, sitim, mille alia taedia, longamque patientiam, promptius iis succurritur, qui facile coguntur, quam quibus inutilis libertas est.* Another tract (ib. 334-40) *de medicina simplicissima et optima, motu, inedia, aquae potu.* Here he cites Plin. XXIV § 5, who laments the importation of foreign drugs, *cum remedia vera pauperrimus quisque cenet.* A tract, *de medico sui ipsius* (239-47) where he says : *miseri sunt illi, qui ea, quae corpori vel prosunt, vel nocent, ipsi non attendunt vel cognoscunt, neque suae naturae scientiam possident, vel diaetae regulas servant, sed pro arbitrio suo vivunt, et persuasum habent medici duntaxat hoc esse officium, cui soli in negotio sanitatis sit fidendum.* Another, *de methodo acquirendi longam vitam* (ib. 247-56) ; another, *de prolonganda litteratorum vita per regulas diaeteticas* (278-88) ; another, containing *septem leges sanitatis* (313-20) ; the seventh is *Fuge medicos et medicamenta, si vis esse salvus.* Celsus (V pr.) commends Asclepiades for preferring the diet cure to drugs.

GEORGE CHEYNE (1671-1743), see for his life the *Biographia Britannica ;* also *Dr. Cheyne's account of himself and of his writings : faithfully extracted from his various works. The second edition.* London, J. Wilfrid, 1743. p. 1 *Upon my coming to London, I, all of a sudden, chang'd my whole manner of living : I found the* bottle-companions, *the* younger gentry, and free-livers *to be the most easy of* access, *and most quickly susceptible of* friendship and acquaintance ; *nothing being necessary for that purpose but to be able to* eat *lustily, and swallow down much liquor.* p. 2 finding his health give way, he gave up suppers, ate but little meat even at dinner, and drank very little fermented liquor. *All my* bouncing, protesting, under-taking *companions forsook me, and dropp'd off like autumn leaves.* p. 21 *I firmly believe, and am as much convinced as I am of any natural effect, that water-drinking only will preserve all the opu-lent healthy from every mortal distemper, bating accidents, heredi-tary and epidemical diseases ; and that a diet of milk and seeds, with water-drinking only, duly continued, and prudently man-aged, with proper evacuation, air and exercise, is the most infalli-ble* antidote *for all the obstinate diseases of the body, and distemper-atures of the mind, so far as it depends on the body, the present state of things will permit.* Cf. pp. 26, 33 verses by E. C.

> *simplex praescripsit medicamen, vivere parvo :*
> *doctrinae exemplar vixit et ipse suae.*
> *nam, dum terra tulit, vinoque et carne cruenta*
> *abstinuit caute, lac erat esca levis.*

p. 37, aphorism I *A constant endeavour after the* lightest *and the* least *of meat and drink a man can be tolerably easy under, is the* shortest *and most infallible* means *to preserve* life, health, *and* serenity. p. 39, aph. 15 *No person of any fortune ever died, or suffer'd* acute *pains or mortal distempers, by the* too cool, too little, *or too insipid in diet: all by the* too hot, high *and* savory : *But virtue and health lie in the* golden mean, *so difficult to be found, and only to be secur'd by the* lightest *and the* least *a man can be tolerably easy under.* aph. 16 *The eternal* law *of nature, by intense pain in* craving *and* hunger, *will never suffer a person in his right senses to go on long obstinately, and to his hurt, in the* too little. aph. 17 Water pure, clear and insipid, *is the sole* beverage *that can procure or continue* health *and a* clear head, *being the* sole *fluid that will pass through the smallest animal tubes without resistance.* pp. 49, 50 *Among us good free-thinking protestants of* England, *abstinence, temperance and moderation (at least in eating), are so far from being thought a* virtue, *or their contrary a* vice, *that it would seem, not eating the fattest and most delicious, and to the* top, were *the only* vice *and disease known among us, against which our parents, relations, friends, and phy-*

sicians exclaim with great vehemence and zeal ; and yet, if we consider the matter attentively, we shall find there is no such danger in abstinence as we imagine ; but, on the contrary, the greatest abstinence and moderation Nature and its eternal laws will suffer us to go into and practise for any time, will neither endanger our health *nor weaken our* just thinking, be it ever so unlimited or unrestrained. It is very observable, that in all civil and established religious worships, hitherto known, among policy'd nations, Lents, days of abstinence, seasons of fasting, and bringing down the brutal part of the rational creature, have had a large share, a strict observance, and been reckon'd an indispensable part of their worship and duty, except among a wrong-headed part of our reformation, where it has been despis'd and ridicul'd into a total neglect ; and yet it seems not only natural, and convenient for health, but strongly commanded in the old and new testament, and might allow time and proper disposition for more serious and weighty purposes : and this Lent, or times of abstinence, is one reason of the chearfulness, or serenity of some Roman Catholic and southern countries, which would be still more healthy and long-liv'd were it not for their excessive use of aromatics and opiates (which are the worst kind of dry drams), and is the cause of their unnatural and unbridled lechery, and shortness of life. For remedying the distempers of the body to make a man live as long as his original frame was design'd to last, with the least pain, fewest diseases, or loss of his senses, I think Pythagoras and Cornaro (for suggesting a general and effectual mean) by far the two greatest men that ever were : the first, by vegetable food and unfermented liquors ; the latter, by the lightest and least of animal food, and naturally fermented liquors. p. 51 I have been credibly inform'd, that Sir Isaac Newton, when he applied himself to what is esteem'd the greatest stretch of human invention and penetration (viz. the study, investigation and analysis of the theory of light and colours) to quicken his faculties, and fix his attention, confin'd himself to a small quantity of bread, during all the time, with a little sack and water, of which, without any regulation, he took as he found a craving, or failure of spirits. pp. 53-54

*People think they cannot possibly subsist on a little* meat, milk *and* vegetables, *or any low diet; and that they must infallibly perish if they should be confin'd to* water *only; not considering that* nine *parts in* ten *of the whole mass of mankind are necessarily confin'd to this* diet, *or pretty nearly to it, and yet live with the use of their* senses, limbs *and* faculties, *without diseases, or but* few, *and those from* accidents *or* epidemical *causes; and that there have been* nations, *and now are numbers of* tribes, *who voluntarily confine themselves to* vegetables *only. . . . . And there are whole villages in this kingdom (even of those who live on the plains) who scarce eat* animal food, *or drink fermented liquors a dozen times in a year.* p. 55 *As to spirits and liquors that have pass'd through the tortures of the fire, they are only of modern invention and* Ottoman *extraction . . . ; and are of such use as the blowing up of an house in an universal conflagration, to save some palace,* viz. *life itself, when in danger. Neither were they ever design'd by Nature and its Author for an animal body, as nourishment or common drink, and scarce deserve a place in the apothecary's shop;* spirits *having made more havoc among mankind, by far, than even gunpowder.*

The Latin edition of Cheyne's *essay of health and long life* is much fuller than the English. *Tractatus de infirmorum sanitate tuenda vitaque producenda.* Lond. 1726. In the preface he declares that no technical knowledge of physiology or anatomy is needed for the application of his rules; that flesh meat and strong drinks may suit those who live by hard labour, but that students require a spare and low diet (cf. pp. 54-56). pp. 3-4 the proverb *misere vivitur, cui medice vivitur.* To be the slave of physic is (as Martial says) *ne moriare mori;* but our duty to God and man commands us to seek health by temperance. pp. 27-98 of meat and drink. pp. 62-66 water the only true drink. Cheyne's *English malady: or, a treatise of nervous diseases of all kinds; as spleen, vapours . . . with the author's own case at large,* sec. ed. Lond. 1734, is also worth reading.

CHRISTOPH WILHELM HUFELAND (1762-1836). His *Makrobiotik* is a classic to this day in France and Germany. The

English translation (somewhat abridged) was reprinted by Erasmus Wilson in 1853. In the preface (27 Oct. 1823) to the fifth
edition he declares : ' Perhaps never did book so entirely spring
from the bottom of the heart as this. Its fundamental idea possessed me from my earliest youth. . . . God has blessed it. I
know from authentic examples that by it many youths have been
kept in the path of virtue and temperance.' He confutes the
apologists of incontinence (II 173-84, Lond. 1797) *Many still
dream of the bad effects of continence.* He gives rules for promoting virtue : *In these two words,* fast *and* labour, *lies the great
talisman against the temptations of this demon.* On the value of
fasting see pp. 296-97, where he cites the case of the actor
Macklin, who died at the age of 99 [read 97]: *when he found
himself ill,* . . . *he always went to bed, took nothing but bread and
water, and by this regimen was generally relieved from every slight
indisposition.*

Many other authorities will be found in James Mackenzie's
History of health, 3rd ed. Edinb. 1760, and in Sir John Sinclair's
Code of health and longevity, 6th ed. Lond. 1829. The 2nd ed.
in 4 vols. is more full in the bibliographical part.

[30] p. 52. *in books.* I would not, if I knew them, name these
books ; but I have myself seen, in a German encyclopædia, ex-
tending to many volumes, a faint dissuasion from incontinence,
followed by a recommendation, if continence is found painful, to
have recourse to prostitutes, 'as the less of two evils.' Dr.
Nichols here, as elsewhere, has done much to promote the physical and moral health of all English-speaking people. In a
brave tract by an Oxford undergraduate (The science of life,
addressed to all members of the universities of Oxford and Cambridge, and to all who are or will be teachers, clergymen, fathers.
Sec. ed. London : J. Burns. 1878) a pamphlet is referred to
(Revelation of quacks and quackery. Baillière, Tindal and Co.)
which 'gives full details of the cruel impositions practised on the
credulous by these quack doctors.'

I cannot refrain from citing a few words from letters of Mr.
Ruskin's, prefixed to ' The Science of Life ': 'indeed there is no

true conqueror of Lust but Love.' 'All that you have advised
and exposed is wisely said and bravely told ; but no advice, no
exposure, will be of use until the right relation exists again between the father and the mother and their son. To deserve his
confidence, to keep it as the chief treasure committed in trust
to them by God : to be, the father his strength, the mother his
sanctification, and both his chosen refuge, through all weakness,
evil, danger, and amazement of his young life.'

[31] p. 52. *family doctors prescribe drunkenness to temperate
ladies.* A perfectly truthful lady told me that her doctor 'ordered' her to drink sherry until she scarcely knew whether she
stood on her head or her heels. I repeated the prescription to
another lady, also perfectly trustworthy, who rejoined : 'My
doctor told me to drink port until I was fuddled.' Surely the
*troisième intermède* of Molière's le malade imaginaire is not too
severe on such professors of the healing art :

> *Vivat, vivat, vivat, vivat, cent fois vivat*
> *Novus doctor, qui tam bene parlat !*
> *Mille, mille annis et manget et bibat,*
> *Et seignet et tuat !*

[32] p. 52. Wordsworth, 'poems dedicated to national inde-
pendence and liberty,' part I, sonnet 13 :

> *O FRIEND ! I know not which way I must look*
> *For comfort, being, as I am, opprest*
> *To think that now our life is only drest*
> *For show ; mean handy-work of craftsman, cook,*
> *Or groom ! We must run glittering like a brook*
> *In the open sunshine, or we are unblest :*
> *The wealthiest man among us is the best :*
> *No grandeur now in nature or in book*
> *Delights us. Rapine, avarice, expense,*
> *This is idolatry ; and these we adore :*
> *Plain living and high thinking are no more :*
> *The homely beauty of the good old cause*
> *Is gone ; our peace, our fearful innocence,*
> *And pure religion breathing household laws.*

[33] p. 53.   George Herbert, 'The church-porch,' st. 23.   On 'sickly healths' I have collected many authorities in the autobiography of Matthew Robinson, Camb. 1856, pp. 46, 112, 210. Add Bishop Brown's Discourse on drinking healths, 1716, 12. Samuel Clarke's Lives (1683) II 126.   Du Soul on Lucian conviv. 16.   Retrosp. rev. XII 322.

[34] p. 53.   Wesley writes to his mother 1 Nov. 1724 (Tyerman I 27) *I suppose you have seen the famous Dr. Cheyne's 'Book of health and long life,' which is, as he says he expected, very much cried down by the physicians.   He refers almost everything to temperance and exercise, and supports most things with physical reasons.   He entirely condemns eating anything salt or high-seasoned, as also pork, fish, and stall-fed cattle; and recommends for drink two pints of water and one of wine in twenty-four hours, with eight ounces of animal and twelve of vegetable food in the same time.   The book is chiefly directed to studious and sedentary persons.*   Wesley wholly left off the use of flesh and wine, and confined himself to a vegetable diet, chiefly rice and biscuit. This he continued during the whole of his residence in Georgia; but on his return to England, for the sake of some who thought he made it a point of conscience, he resumed his former mode of living, and practised it to the end of life, except during a two years' interim, when he again became vegetarian and teetotaller, because Dr. Cheyne assured him that it was the only way to be free from fevers (Tyerman I 117, cf. 525).   See his arguments against tea-drinking (ibid. 521-3).   In his journals, on his birthdays (28 June), there are many valuable notices on his health. Thus in 1786 *I am entered into the eighty-third year of my age. I am a wonder to myself.   It is now twelve years since I have felt any such sensation as weariness.   I am never tired (such is the goodness of God!) either with writing, preaching, or travelling; one natural cause undoubtedly is my continual exercise and change of air.*   See also under 1774, 1782, 1788.   His *primitive physick; or, an easy and natural method of curing most diseases,* went through twenty-three editions in his life-time.

[35] p. 53.   *the Old Catholic churches.*   In the first synod of the

German Old Catholics at Bonn, confession and fasting—great instruments of tyranny in the Vatican communion—were left to the discretion of each member of the church. Fasting, as enjoined by the bishops, imposed an intolerable burden on the poor, while to the rich it only offered an agreeable change of fare. [It costs more *faire maigre* at a hotel than *faire gras !*] In no case ought fasting to be allowed to prejudice the health.

In the Deutscher Merkur, 3 April 1880, is a pastoral of bishop Reinkens, explaining why he never issued a 'Fasten-Hirten-brief.' He had received from no 'Lord' a commission to give directions respecting the difference of meats. The only true Lord of the church has laid down the authoritative rule (Luke x 8) *into whatsoever city ye enter, eat such things as are set before you.* The reason is given by St. Paul, Rom. xiv 14 17, Col. ii 16. The bishop also cites 1 Cor. viii 8-9, Rom. xiv 3 22-23, 1 Tim. iv 3, Matt. vi 16-18.

[36] p. 54. *plures gula quam gladius.* See Erasmus (adag. 468-9, ed. Francof. 1656) under *gula plures quam gladius perimit*, who cites the Fr. *Gourmandise tue plus de gens que l'espée.* Franc. Patric. de republica v 8 *gula plures occidit quam gladius estque fomes omnium malorum.* Lessius (p. 105, ed. Camb. 1634) cites *non plures gladio quam cecidere gula.*

[37] p. 54. Seneca cites as a rule of great men ep. 18 § 5 *interponas aliquot dies, quibus contentus minimo ac vilissimo cibo, dura atque horrida veste, dicas tibi* 'hoc est quod timebatur.' § 6 some imitated poverty every month, reducing themselves almost to want, that they might never fear what they had often learnt. § 7 *non est quod existimes me dicere Timoneas cenas et pauperum cellas et quicquid aliud est per quod luxuria divitiarum taedio ludit : grabatus ille verus sit et sagum et panis durus ac sordidus. hoc triduo et quatriduo fer, interdum pluribus diebus, ut non lusus sit, sed experimentum . . . . § 8 non est tamen, quare tu multum tibi facere videaris. facies enim, quod multa milia servorum, multa milia pauperum faciunt : illo nomine te suspice, quod facies non coactus, quod tam facile erit tibi illud pati semper quam aliquando experiri . . . § 9 certos habebat dies ille magister*

*voluptatis Epicurus, quibus maligne famem extingueret, visurus,
an aliquid desset ex plena et consummata voluptate, vel quantum
desset, et an dignum, quod quis magno labore pensaret.  hoc certe
in his epistulis ait, quas scripsit. . . . ad Polyaenum : et quidem
gloriatur non toto asse pasci : Metrodorum, qui nondum tantum
profecerit, toto.*  § 10 *in hoc tu victu saturitatem putas esse?  et
voluptas est.  voluptas autem non illa levis et fugax et subinde
reficienda, sed stabilis et certa.  non enim iucunda res est aqua et
polenta aut frustum hordeacei panis, sed summa voluptas est posse
capere etiam ex his voluptatem et ad id se deduxisse, quod eripere
nulla fortunae iniquitas possit.*  § 11 *liberaliora alimenta sunt
carceris.*  See some excellent remarks on periodical fasts in Plut.
II 135*d* (de sanitate, 23.  This instructive tract is translated in
Sinclair's Code of health, ed. 2, and deserves to be reprinted by
some sanitary association.)

[38] p. 55.   Xen. mem. I 3 § 6.   Cf. Wyttenbach on Plut. II
124*d*, Gataker advers. posth. XLIII p. 882*g*, Clem. Al. paed. II
1 § 15.

[39] p. 55.   Sen. ep. 21 § 10 when you enter the gardens of
Epicurus and read the inscription : HOSPES, HIC BENE MANEBIS,
HIC SUMMUM BONUM VOLUPTAS EST : *paratus erit istius domi-
cilii custos hospitalis humanus et te polenta excipiet et aquam
quoque large ministrabit et dicet : ' ecquid bene acceptus es?  non
irritant' inquit ' hi hortuli famem, sed exstinguunt.  nec maiorem
ipsis potionibus sitim faciunt, sed naturali et gratuito remedio
sedant.  in hac voluptate consenui.'*  § 11 *de his tecum desideriis
loquor, quae consolationem non recipiunt, quibus dandum est aliquid, ut desinant : nam de illis extraordinariis, quae licet differre,
licet castigare et opprimere, hoc unum commonefaciam : ista voluptas naturalis est, necessaria.  huic nihil debes : si quid impendis,
voluntarium est.  venter praecepta non audit.  poscit, appellat*
('duns').  *non est tamen molestus creditor : parvo dimittitur, si
modo das illi quod debes, non quod potes.*  ibid. 60 § 3 *quantulum
est enim, quod naturae datur !  parvo illa dimittitur.  non fames
nobis ventris nostri magno constat, sed ambitio.*   See my notes on
Iuv. XIV 318-20.  Compare the excellent translation of the rules

of the school of Salerno : *L'école de Salerne, traduction en vers français par Ch. Meux Saint-Marc.* Paris, Baillière, 1880. On the verses (p. 78)

> *tu numquam comedas, stomachum nisi noveris esse*
> *purgatum vacuumque cibi quem sumpseris ante ;*
> *ex desiderio poteris cognoscere certo,*

we read (pp. 301-2) : *L'appétit, tel devrait être le seul régulateur des repas. Ne mangez que quand vous sentez le désir de le faire ; en suivant cette simple formule, vous éviterez des causes innombrables de maladies. Toute l'hygiène gastronomique est contenue dans ces excellents préceptes ;* malheureusement, il n'en est pas qui soit moins suivi. L'homme civilisé s'est fait en toute sorte de choses des appétits factices, et ils ne sont pas les moins impérieux. Il confond une soif légère et l'appétit avec la soif et la faim, qui sont des besoins qu'il importe de satisfaire. Il y a trois sortes d'appétits :... L'appétit vient en mangeant.... Il est vrai que l'homme peut pervertir à la longue les sensations involontaires. En général, dans les classes aisées, l'homme mange trop. La bonne règle, suivant Hippocrate, est de quitter la table avant la satiété. On digère bien ce que l'on mange avec appétit, parce que l'estomac alors se trouve vide et que la diète est le meilleur remède contre la plénitude de cet organe et contre l'indigestion. Il ne s'agit pas d'ingérer des aliments dans l'estomac, mais de les digérer. Toutes les fois que la digestion des aliments que l'estomac a déjà reçus n'est pas complète, il faut s'abstenir d'en prendre de nouveaux..... Seul, dans la nature, l'homme connaît l'obésité, ce châtiment des gros mangeurs, et la goutte, cette expiation des gourmands.* Cheyne, *The English Malady* (1734) 299 *A* vegetable patient *of mine very justly observ'd to me, That whereas before, he could never trust his appetite's longings or craving, while on an* animal high diet, *without suffering to extremity ; now he found he might safely and securely trust nature and appetite,* WITHOUT DANGER, *fear, or suffering.* See Hufeland, *Art of prolonging life*, London 1797, II 44-45 *consumendo consumimur.*

⁴⁰ p. 55. See Dr. Murchison in the Contemporary Review XXXIV 137 *to a third and by no means a small class of persons,*

*alcohol, even in small quantities, is an unmistakable poison.* p.
139 *A man who is in good health does not require it, and is pro-*
*bably better without it ;* ... *its habitual use, even in moderation,*
*may and often does induce disease gradually.* Dr. Johnson (Boswell, May 1776, p. 513a of Croker's 1 vol. ed.) says : *Every man*
*is to take existence on the terms on which it is given to him. To*
*some men it is given on condition of not taking liberties, which*
*other men may take without much harm. One may drink wine,*
*and be nothing the worse for it : on another, wine may have*
*effects so inflammatory as to injure him both in body and mind,*
*and perhaps make him commit something for which he may de-*
*serve to be hanged.* Galen, de sanitate tuenda V 12 (VI 377 K)
εἴ τις δ' ἄρα, ὡς ἐν ὑγιεινῇ δυσκρασίᾳ, θερμοτάτη κρᾶσις ᾖ, ταύτῃ
συμφέρει μηδ' ὅλως οἶνον διδόναι. ib. 11 (p. 364) the utter diver-
sity of medical opinion about wine. ib. 5 pr. p. 334 wine most
hurtful to boys, most useful to old men.

In a letter to the abp. of Canterbury (10 Oct. 1873 in House
and Home 12 Apr. 1879) Sir Henry Thompson says : *There is*
*no greater cause of evil, moral and physical, in this country than*
*the use of alcoholic beverages. I do not mean by this that extreme*
*indulgence which produces drunkenness. The habitual use of*
*fermented liquors to an extent far short of what is necessary to*
*produce that condition, and such as is quite common in all ranks*
*of society, injures the body and diminishes the mental powers to an*
*extent which I think few people are aware of. Such, at all*
*events, is the result of observation during more than twenty years*
*of professional life devoted to hospital practice, and to private*
*practice in every rank above it. Thus, I have no hesitation in*
*attributing a very large proportion of some of the most painful and*
*dangerous maladies which come under my notice, as well as those*
*which every medical man has to treat, to the ordinary and daily*
*use of fermented drink taken in the quantity which is conventionally*
*deemed moderate.* In a speech at Exeter Hall, 7 Febr. 1877,
before the National Temperance League (ibid.) Sir Henry said :
*Of all the people I know who cannot stand alcohol, it is the brain-*
*workers.* Again in articles reprinted from the Nineteenth Cen-

tury for June and July 1879 (House and Home 26 July 1879) :
*I am of opinion that the habitual use of wine, beer, or spirits is a
dietetic error, say, for nineteen persons out of twenty. In other
words, the great majority of the people, at any age, or of either sex,
will enjoy better health, both of body and mind, and will live
longer, without any alcoholic drinks whatever, than with habitual
indulgence in their use, even although such use be what is popularly
understood as moderate. . . . No one probably is any better for
tobacco ; and some people are undoubtedly injured by it ; while
others find it absolutely poisonous, and cannot inhale even a small
quantity of the smoke without instantly feeling sick or ill. And
some few indulge the moderate use of tobacco all their lives without
any effects, at all events that are perceptible to themselves or to
others.* Even Barrow hailed tobacco as a *panacea*, but experience
(*stultorum magistra*) has proved that *nicotine* is one of the
deadliest of poisons. Yet how many of our students ruin their
health and fortunes by indulgence in this filthy drug. Oxford
University Commission Report (1852) p. 24 : *We cannot forbear
from alluding also to the excessive habit of smoking, which is now
prevalent. Tobacconists' bills have, and that not in solitary in-
stances, amounted to £40 a year.*

[41] p. 55. A lady said : *Dr. Johnson, do take a little wine ; it will
do you good.* 'My dear madam, I cannot take a little.' I cannot
at this moment lay my hand on the passage (it is not, I think, in
Boswell), but see Croker's index under 'drinking' and 'wine'
for the sentiment. Thus 16 Sept. 1773 (p. 336) : *he was prevailed with to drink a little brandy* [for a cold] *when he was going
to bed. He has great virtue in not drinking wine or any fermented liquor, because, as he acknowledged to us, he could not do
it in moderation. Lady Macleod would hardly believe him, and
said, 'I am sure, sir, you would not carry it too far.'* JOHNSON :
*Nay, madam, it carried me. I took the opportunity of a long
illness to leave it off. It was then prescribed to me not to drink
wine ; and having broken off the habit, I have never returned to
it.* 16 March 1776 (p. 480) *Finding him still persevere in his
abstinence from wine, I ventured to speak to him of it.* JOHNSON :

*Sir, I have no objection to a man's drinking wine, if he can do it in moderation. I found myself apt to go to excess in it, and therefore, having been for some time without it on account of illness, I thought it better not to return to it. Every man is to judge for himself, according to the effects which he experiences. One of the fathers tells us, he found fasting made him so peevish that he did not practise it.* Sept. 1777 (p. 551) *Dr. Johnson recommended to me, as he had often done, to drink water only: 'For,' said he, 'you are then sure not to get drunk; whereas, if you drink wine, you are never sure.' I said drinking wine was a pleasure which I was unwilling to give up. 'Why, sir,' said he, 'there is no doubt that not to drink wine is a great deduction from life; but it may be necessary.'* 7 April 1778 (p. 578). ib. p. 579*b Talking of a man's resolving to deny himself the use of wine from moral and religious considerations, he said: 'He must not doubt about it. When one doubts as to pleasure, we know what will be the conclusion. I now no more think of drinking wine than a horse does. The wine upon the table is no more for me than for the dog who is under the table.'* March 20 1781 (p. 678) *The first evening that I was with him at Thrale's, I observed that he poured a large quantity of it* [wine] *into a glass, and swallowed it greedily. Many a day did he fast, many a year did he refrain from wine: but when he did eat, it was voraciously; when he did drink wine, it was copiously. He could practise abstinence, but not temperance.* The same was the case with our late registrary, Mr. Romilly.

[42] p. 55. Sall. Catil. 52 § 11 *iam pridem equidem nos vera vocabula rerum amisimus: quia bona aliena largiri liberalitas ... vocatur.*

[43] p. 56. Meadley's Life of Paley, Sunderland 1809, pp. 193-94 'on the authority and in the very words of a gentleman who was present' when Paley made the statement at Cambridge in 1795.

[44] p. 57. Tert. virg. vel. 1 *sed dominus noster Christus veritatem se, non consuetudinem nominavit. ... hunc qui receperunt, veritatem consuetudini anteponunt.* Cypr. ep. 71 3, 73 13, 74 9

*consuetudo sine veritate vetustas erroris est.* 75 19 sentent. epis-
coporum 63 and 77 (in Hartel's Cypr. pp. 456, 458). Aug. de
bapt. c. Donat. III §§ 9 12, IV § 7, VI §§ 68 71-2. Cf. Wordsworth (ode on the intimations of immortality in childhood) :
> ' Full soon thy soul shall have her earthly freight,
> And custom lie upon thee with a weight,
> Heavy as frost, and deep almost as life !'

[45] p. 58. Iuv. III 182-3 n. *hic vivimus ambitiosa paupertate
omnes.*

[46] p. 58. *Aschams.* See the elaborate German life of Roger
Ascham by Dr. Alfred Katterfeld, Trübner, London 1879. Both
he and Cheke received pecuniary help from Dr. Nic. Metcalfe
(Ascham, scholemaster, 1863, 160-1). On 21 Nov. 1547, Ascham
penned, in the name of the college, a letter to the protector
Somerset (Aschami epist. ed. 1703, p. 292) in which he says that
some members of the college received 3*d.* a week, some 7*d.*, the
most fortunate 12*d.* for commons. Compare Thos. Lever's
description of the slender dinner at 10 a.m. (Baker, Hist. of
St. John's college, p. 131) 'where as they be contente wyth a penye
pyece of biefe amongest iiii, hauynge a fewe porage made of the
brothe of the same byefe, with salte and otemel, and nothynge
els. . . . [At 5 p.m.] they have a supper not much better then
theyr diner.' Cf. Notes and Queries 5 S. III 266 : Thos. Cogan,
Haven of health, 1586 (? -9) in his time at Oxford 'they used
commonly at dinner boyled biefe with pottage, bread and beere,
and no more. The quantity of biefe was in value an halfepenny
for one man, and sometimes, if hunger constrained, they would
double their commons.' Dr. Katterfeld naturally exclaims (p.
25) : 'What a difference between this picture and the rich, com-
fortable luxury which nowadays everywhere meets the visitor in
Cambridge !'

[47] p. 58. *Lady Margaret's name a talisman against luxury.*
See C. H. Cooper's Memoir of Margaret, countess of Richmond
and Derby, Camb. 1874. Bp. Fisher, mornynge remembraunce
had at the moneth mynde of the noble princes Margarete
(Fisher's English Works, E. E. T. S. I 293-4, Lond. 1876) : 'her

sobre temperaunce in metes and drynkes was knowen to al them that were conversaunt with her, where in she lay in as grete wayte of herself as ony persone myght, kepinge alway her strayte mesure, and offendyng as lytel as ony creature myght. Eschewynge bankettes, reresoupers, ioncryes betwyxe meales. As for fastynge, for aege and feblenes albeit she were not bounde, yet tho dayes that by the chirche were appoynted she kept them diligently and sereously, and in especyall the holy lent, thrughout that she restrayned her appetyte tyl one mele and tyl one fysshe on the day, besyde her other peculer fastes of deuocion, as saint Anthony, Mary Maudeleyn, saynt Katheryn with other. And thorowe out al the yere the fryday and saterday she full truely obser    d.'

[48] p. 58. *athletics a help, not a hindrance.* See Mr. Lyttelton's well-timed caution in the Nineteenth Century for Jan. 1880.

[49] p. 58. *George Selwyn.* In Tucker's life of Bp. Selwyn many illustrations of his simplicity of tastes will be found. Thus, when out on a long walk in England, he would lunch on apples and a crust of bread, and sup on macaroni at night.

---

*Temperantia est moderatio cupiditatum rationi oboediens*
Cic. *fin.* ii § 60.

*Non est, non (mihi crede) tantum ab hostibus armatis aetati nostrae periculum, quantum ab circumfusis undique voluptatibus. qui eas temperantia sua frenavit ac domuit, multo maius decus maioremque victoriam sibi peperit, quam nos Syphace victo habemus*
Scipio in Liv. xxx 14 §§ 6, 7.

www.ingramcontent.com/pod-product-compliance
Lightning Source LLC
Chambersburg PA
CBHW021953190326
41519CB00009B/1242